SpringerBriefs in Cybersecurity

Cybersecurity is a difficult and complex field. The technical, political and legal questions surrounding it are complicated, often stretching a spectrum of diverse technologies, varying legal bodies, different political ideas and responsibilities. Cybersecurity is intrinsically interdisciplinary, and most activities in one field immediately affect the others. Technologies and techniques, strategies and tactics, motives and ideologies, rules and laws, institutions and industries, power and money—all of these topics have a role to play in cybersecurity, and all of these are tightly interwoven.

The SpringerBriefs in Cybersecurity series is comprised of two types of briefs: topic- and country-specific briefs. Topic-specific briefs strive to provide a comprehensive coverage of the whole range of topics surrounding cybersecurity, combining whenever possible legal, ethical, social, political and technical issues. Authors with diverse backgrounds explain their motivation, their mindset, and their approach to the topic, to illuminate its theoretical foundations, the practical nuts and bolts and its past, present and future. Country-specific briefs cover national perceptions and strategies, with officials and national authorities explaining the background, the leading thoughts and interests behind the official statements, to foster a more informed international dialogue.

More information about this series at http://www.springer.com/series/10634

Sanjay Goel · Yuan Hong
Vagelis Papakonstantinou
Dariusz Kloza

Smart Grid Security

Sanjay Goel
Department Information Technology
  Management
University at Albany
Albany, NY
USA

Yuan Hong
Department of Information Technology
  Management
University at Albany
Albany, NY
USA

Vagelis Papakonstantinou
Research Group on Law, Science,
  Technology & Society (LSTS)
Vrije Universiteit Brussel
Brussels
Belgium

Dariusz Kloza
Research Group on Law, Science,
  Technology & Society (LSTS)
Vrije Universiteit Brussel
Brussels
Belgium

ISSN 2193-973X   ISSN 2193-9748   (electronic)
SpringerBriefs in Cybersecurity
ISBN 978-1-4471-6662-7   ISBN 978-1-4471-6663-4   (eBook)
DOI 10.1007/978-1-4471-6663-4

Library of Congress Control Number: 2015932438

Springer London Heidelberg New York Dordrecht

Printed on acid-free paper

Springer-Verlag London Ltd. is part of Springer Science+Business Media
(www.springer.com)

Foreword

An irrevocable process has been set into motion: The world is growing smart. Every kind of technology is presently reconsidered as transformable into an intelligent, sensory and communicative, networked device. There are many good reasons for this reconsideration. As has been seen in the past, such transformations have many benefits. Technological processes can be of higher density, better synchronized, of higher efficiency, less prone to human error, they can offer more functionally and quite generally become more profitable. In sum, many stakeholders emerge to design such technologies and initiate a market for products.

But as ever so often with technology in general and information technology in particular, there are risks and side effects. In this current paradigm of "smartification" of our previously "dumb" technologies, the risks of those old technologies are confronted with and infiltrated by the functionalities and risks of the new technologies. This fusion challenges many initial assumptions and established concepts for safety and security in novel ways. Old risks such as physical damages of many of our old technologies, previously forged into acceptable states by electromechanical safety concepts, may reemerge in different shapes and sizes when chips and logic replace single switches and valves, and when the net is slowly creeping in. A smart car or a smart factory may be more efficient and more transparent to some extent, but it will also be more open to outsiders, more accessible, it might be more prone to unwanted complex developments, as any kind of IT always adds tremendous complexity, and it will most certainly require much more attention, more maintenance and more expertise. Also, entirely new risks may come up, such as privacy concerns simply by driving a car or heating the house, given the fact that "smart" technologies generate data—which can be information about people to some extent.

Accordingly, this new fusion of old technologies and new ones requires foresight and wisdom. It may in fact already be a little to late to call for that. The engineers and industries have started these paradigms a long time ago, and approach implementation and sales fast now. Too much money and effort has been spent to pause and reconsider everything. This, by the way, is a very classical problem of technology research. As long as a technology is in its infancy, its actual impact and its use models cannot be predicted with high certainty, so its risks and side effects are difficult to pinpoint. The technology researcher can only guess and hypothesize, which in turn renders much of her effort into an ivory tower perspective. Only once technologies reach a first stage of maturity, with use models in actual implementation, more precise, correct and relevant assumptions can be made. But then, too much money has been spent and too many paradigms are in implementation already to return to the drafting board and start over on some fundamentals. At that point, technical and economical path dependencies are established. They can be reformed, to be sure. The widely known saying that you cannot stop progress is a little imprecise to this end. Progress at large may be unstoppable for many reasons. But any particular progress can always be shaped and directed, even reconsidered and revoked entirely, if its benefits are not nearly in line with its risks—especially as long as it is not too established yet.

Smart technologies, fortunately, are still comparatively young and could be viewed with a kind of "design optimism". They are dangerous, to be sure, risky and difficult, as two highly complex types of technologies are melted into each other, with a lot of difficult scenarios emerging. But they should also be considered malleable and even an opportunity. Any fresh start in innovation is also a chance to do things better this time. Information technology is so incredibly bad in its security, so open to sabotage and espionage, to surveillance and manipulation that a reform within an environment with much higher concerns in safety and security could force it to return to some of its fundamental issues and try harder.

A precondition for such an effort is a thorough understanding and a good and causally well-defined structuring of the problems and their roots. They have to be intelligible in their technical, economic, legal and societal dimensions, so options and opportunities can be developed and recommended for implementation. To this end, the whole process of "smartification" still requires a lot more literature, especially interdisciplinary writings, connecting the technical and the human world and reflecting the possible realities of smart worlds.

This SpringerBrief aims to fill this gap in the important field of smart power. Smart power (also called "smart grids") is a first larger technical area under reform by networked information technology. It is already in application and implementation and can be assessed in its processes and regulations, its technicalities and risks.

The authors of this brief have done an excellent and outstanding job illuminating this new field and explaining the risks and benefits, the conditions and opportunities, causalities and first actors of this field. As a result of their great work, this brief will serve as an excellent guide—a true briefing—not just to smart power, but to the whole emerging smart world and its core topics.

ESMT Berlin, January 2015   Dr. Sandro Gaycken

Contents

Chapter 1
Security Challenges in Smart Grid Implementation

Sanjay Goel and Yuan Hong

Abstract The smart grid architecture amalgamates the physical power grid and a communication grid into a single monolithic network. It poses several security threats that are well known (Li et al. in IEEE Trans Smart Grid 3:1540–1551, 2012 [1], McDaniel and McLaughlin in IEEE Secur Priv 7:75, 77, 2009 [2], Bisoi and Dash 2011 [3]). However, it faces unknown threats from the cyber-physical interfaces whereby either cyber-threats can lead to actuation of physical devices or vice versa if physical devices could be manipulated to disrupt the communication infrastructure. The most prevalent threats to the operation and safety of the smart grid come from physical destruction of infrastructure, data poisoning, denial of services, malware, and intrusion. The most prevalent threat to the consumer is breach of privacy of the data and malicious control of personal devices and appliances. This chapter articulates the smart grid architecture and the cyber-physical threats to which the smart grid is vulnerable.

1.1 Smart Grid Architecture

1.1.1 Introduction

The smart grid is a traditional power grid with a communication network overlaid on top of the traditional power grid. The communication and power grid are interrelated such that the communication network depends on the power grid for data and the power grid depends on the communication for operational activities. The role of the grid is to provide ubiquitous communication capability for collecting data from sensors and meters, process it in situ, and provide pertinent information to support multiple activities such as ensuring grid stability, detecting and resolving anomalies, forecasting load, and facilitating demand response. All this needs to be done while protecting the privacy of the consumers, protecting critical operational data that from national adversaries, and ensuring the integrity of the data for both business and operational needs. This is not a trivial challenge for several reasons, including need

© The Author(s) 2015
S. Goel et al., *Smart Grid Security*, SpringerBriefs in Cybersecurity,
DOI 10.1007/978-1-4471-6663-4_1

1

to integrate disparate communication media into a single monolithic network, need to provide guaranteed latency and bandwidth for several applications, and need to ensure privacy and security of the data as necessary.

The power grid is typically segregated into transmission, distribution, and the last mile. Transmission carries high-voltage current over long distances to substations. Distribution carries lower-voltage data from substations to local transformers. The last mile connects the local transformers to consumers, and it is where utilities and consumers interact to support real-time management of energy generation, distribution, usage, and efficiency. With the integration of the smart grid technologies, the traditional network is now entering households and businesses. Parallel to the power grid, the communication grid can be segregated into wide area network (WAN), metropolitan area network (MAN), field area network (FAN), and home area network (HAN) as shown in Fig. 1.1.

The primary goal associated with the transmission network is to provide situational awareness where technologies for monitoring and control of the grid across a large geographical network are necessary. This will include incorporation of synchrophasors for monitoring the state of the grid to ensure its synchronization as well as supporting SCADA systems. Any failure at this level will have far-reaching consequences on the stability of the entire grid including large-scale blackouts. Consequently, WAN will need to provide high bandwidth (600–1500 kbps), low latency (20–200 ms), and high reliability (over 99.999 %). This kind of reliability will probably not be met by wireless technologies and will rely primarily on fibre optic or other wired technology. At the distribution level, the goal is to be able to monitor the distribution network for faults and other anomalies as well as to be able to integrate microgeneration sources. This will have a variable requirement for bandwidth (10–100 kbps) and latency (from 10 ms to 15 s) with a reliability greater than 99 %.

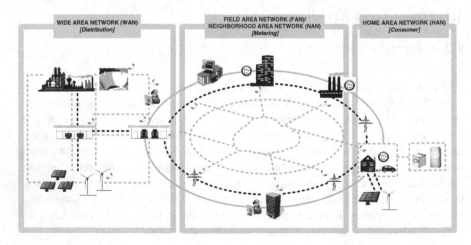

Fig. 1.1 The evolution towards the smart grid

A key requirement would be to handle peak data from multiple sources during power outages. These networks are typically dense and entrenched throughout the city requiring a combination of different technologies including wireless, PLC, and AMI. The last mile would be responsible for metering from the customer as well as providing demand response capability. This would require vendor interoperability to be able to support different types of devices in customer homes. Redundancy, fault tolerance, and security are all critical for this network. HAN would require a short range with the ability to penetrate through walls with very high data rate from multiple appliances. The communication channel should be able to handle a barrage of interference from multiple devices and be able to operate reliably. For aesthetics and convenience, the HAN network will most likely be wireless.

1.1.2 Communication Technologies

Currently, most power system infrastructure uses a combination of multiple technologies including dedicated cable, microwave, power line communication, and fibre optic technology. Replacing all existing infrastructure with dedicated fibre optic communication would be cost prohibitive. The infrastructure consequently would be a combination of wireless, fibre optic, power line carrier (PLC), and traditional cable or Ethernet.

One of the most seductive technologies to implement would be the PLC given that the power infrastructure already connects together the entire grid across all levels of the grid. The technology has been developed since 1920, initially for voice and data communication over high-voltage lines between remote stations and most recently for load control and automatic meter reading. The earlier technology was very narrow band operating below 3 kHz frequency resulting in low data rate of 60 bps which could be transmitted over large distances. The CENELEC standard in 1992 regulates the use of spectrum in four bands: 3–95 kHz for power utilities; 95–125 kHz for general applications, 125–140 kHz for home networks, and 140–148.5 for security applications. An innovation that is propagating at the WAN level is the use of optical fibres encased in the ground wire that runs on top of all the transmission towers to take a preferential lightning hit. Most electrical power grid systems in the world use the ground wire with an optical fibre encased in it. These communication channels operate efficiently over large distances with minimal losses and high reliability. These optical fibres in the newer installation of transmission lines facilitate the deployment of smart grid without any need for additional communication capacity. While such an infrastructure supports the requirements of the smart grid, the TCP/IP protocol that drive communication today would not provide the requisite security required for communication among power plants (including nuclear), control equipment, substations, and eventually distribution grids.

Wireless media will be a critical part of the smart grid communication infrastructure primarily due to convenience and accessibility especially in the area of metering and home area networking. Communication is possible by transmitting from hop to hop (electrical poles) across large distances. There are several different technologies that can be used including Microwave, WIMAX, MESH, LTE, Cellular, WLAN, and Zigbee. Microwave is a high-capacity point-to-point wireless transport for providing a backbone to telecommunication services including radio access network and WAN. It can be used for applications such as SCADA, AMI, and Demand Response. WiMAX is a cost-effective channel broadband connectivity across large areas as an alternative to GSM and CDMA. It can be used for AMI, SCADA, demand response, mobile workforce, and video surveillance. Mesh network is created by using a network of radio nodes arranged in a mesh topology and is commonly used for providing the last mile of connectivity for broadband access. It can overlay or replace copper DSL or provide a redundant channel of communication. It can be used for remote monitoring, demand response, AMI, and distribution automation. The problem is delay caused due to hops from router to router; however, it is easily expandable by adding additional nodes and permits building redundancy in the network. LTE is the next-generation network for mobile communication that provides high spectral efficiency and low latency. It can be used for all applications in which mesh network are used; however, it is not readily available and cost of installation is high. Cellular networks are typically used for most consumer applications including mobile phones, Internet connectivity, voice and video chat, and text messaging. It can be used in the smart grid for workforce coordination, AMI, etc. The main advantage is that it is already widely deployed requiring minimal capital costs for operationalizing smart grid initiatives. Wireless LAN (WLAN) is already used extensively for indoor connectivity and could be leveraged easily for home area networking and connecting smart meters with internal visualization devices. Zigbee is a standard developed specifically for the smart grid targeted at networking in-home applications including smart meters, smart lighting, and appliances.

1.1.3 Sensors and Devices

While the communication infrastructure is the enabler for the smart grid, the real benefit will come from the sensors and devices on the network. A smart meter will be installed on each node of the network that will facilitate a two-way exchange of power through metering in both directions and allow fine-grained control of electricity usage of customer appliances to the utility company. The meters will also allow remote access to appliances in the households to customers and provide them with detailed usage statistics. In addition, it will provide commercial entity access to devices for monitoring, diagnostics, and repair. The smart grid will also minimize the manual data collection from the grid.

Until recently, utility company employees have manually gathered operational data including electricity metering, identifying broken equipment, and faults. The smart grid infrastructure will allow remote control and automation of several operational activities including monitoring of the distributed infrastructure comprised of wires, substations, transformers, switches, etc. Each device on the network will contain sensors to gather data (voltage, phase, temperature, etc.). That data will be relayed to the control centre through the two-way communication system of the grid. One of the key needs of the grid is improved stability that will require synchronized phasor (synchrophasor) devices installed throughout the network for data collection. Synchrophasors will provide real-time measures of electrical quantities from the entire power grid for several critical applications including estimation of dynamic state response, grid synchronization, and fault identification. These devices consist of GPS satellite-synchronized clocks, phasor measurement units (PMUs), phasor data concentrator, and analysis software.

Another key element of the smart grid is the self-healing of the grid that can correct flaws automatically or isolate the faults to minimize the outages for consumers. To develop self-healing abilities in the grid, a processer will be required in each switch, and circuit breaker and electromechanical switches will need to be replaced with solid-state electronic circuits. Automated reclosers will be added on to the grid to allow temporary instantaneous faults caused by events such as falling tree limbs and heavy winds to be self-corrected. To manage and analyse the data, distributed analytic processing capability as well as storage in the grid will need to be incorporated. Finally, the grid will need to be secured both through perimeter defence and improved visibility into the network for intrusions and attacks as we will discuss that further.

Summary

Smart grid requires a massive communication infrastructure with complete connectivity across the entire country. Based on geographically dispersed infrastructure elements, communication will need to be a hybrid with a variety of communication media. Initially, communication will be shared by other services at least in the distribution network; however, over time, communication networks are likely to become more dedicated as communication infrastructure is laid out exclusively for the smart grid. The grid infrastructure also requires sensors for monitoring and diagnostics throughout the grid as well as upgrading the existing electromechanical switches to electronic switches for imbuing self-correction ability in the grid. A key imperative to the success of the smart grid would be a robust security mechanism that not only prevents intrusion but also ensures privacy of customers and integrity of the data.

1.2 Smart Grid Security Concerns and Threats

The smart grid is poised to fundamentally change the electrical grid from the centralized utility-centric grid to a distributed consumer-centric grid where the consumers are well informed and active participants in energy consumption and

generation. The smart grid also brings improved visibility into the grid that will help
in better monitoring and control of the grid to ensure stability and reduce chances of
large-scale blackouts. A ubiquitous communication network that connects all the
users, utilities, and producers into a monolithic network enables this functionality.
However, all this comes at a cost, which is increased risk of cyber-attacks. There are
threats from several actors including terrorists, nation states, criminals, and disgruntled employees. In addition, there is need to protect customer privacy which
can be revealed through the fine-grained transmission of usage data. If security is in
adequate, the communication network in the grid can become a liability rather than
an asset. There have been numerous attacks on the smart grid, and there are several
security threats, some of which we discuss in this chapter.

1.2.1 Reported Attacks on Electric Grids

There have been several documented impact on the electric grid attributed to targeted cyber-attacks or as unintended consequences of network anomalies that led to
SCADA system failures [4] described ahead. In January 2003, the Slammer worm
infected a computer network at the Davis-Besse nuclear power plant in Oak Harbor,
Ohio, disabling a safety monitoring system and the plants process computer for
several hours. In August 2003, a failure of the alarm processor of FirstEnergy
prevented monitoring of the grid and as several transmission lines tripped for
various reasons, a cascading failure resulted in disabling power plants through
north-east and leading to an extended blackout. In August 2006, circulation pumps
at the Brown Ferry nuclear plant in Alabama failed due to excessive traffic on the
control system network. Investigation of a 2009 incident revealed that hackers were
able to steal power by hacking into smart meters and changing the power consumption reading. Phishing incidents were also detected at an electric bulk provider
and malware samples were detected that indicated a targeted and sophisticated
intrusion.

Most of the above attacks raised concern, and there has been innuendo regarding
the participation of nation states in these attacks. There were also attacks that were
in the category of information warfare and propaganda such as the attack on Estonia
and Georgia during conflicts with Russia. The first major cyber-warfare attack that
attacked the critical infrastructure of a country was the Stuxnet attack that was
targeted at degrading the Iranian nuclear enrichment facilities. Stuxnet is a worm
that exploits multiple zero-day vulnerabilities that make use of stolen digital cer-
tificates to control WinCC SCADA application on Siemens S7 PLC Microcon-
troller [5]. The payload for Stuxnet was delivered using infected USB drives of
nuclear inspectors. The malware was not only able to increase the RPM of the
centrifuges used for enriching uranium, but it also made it appear that the centrifuges were operating normally. This was the first major strategic attack on critical
infrastructure of another country and had propelled countries into an arms race to
develop such weapons as strategic options both for deterrence and counter-attacks.

There have been several data for reconnaissance and probing the critical infrastructure [5]. Night Dragon was an intrusion ostensibly originating in China [6] and aimed at probing industrial control systems of energy companies (oil, gas, and petrochemical) in the United States. The attacks used a combination of social engineering and vulnerabilities in remote administration tools on Windows platforms to break into critical computers on the network to gather proprietary information including documents related to oil and gas field exploration and business negotiations as well as details of SCADA systems. Researchers in Budapest discovered another computer malware named Duqu which is a collection of tools and services including keystroke loggers, kernel drivers, and injection tools. It was found on computers in companies manufacturing industrial control systems. There is speculation that the malware was used by Stuxnet writers to collect information that went into development of the Stuxnet. An even more sophisticated malware targeted at control systems was the Flame toolkit which includes a backdoor, Trojan, as well as replicator and propagation mechanism that allows it to propagate on the network and removable media. Flame is an intelligence-gathering malware that can sniff traffic, take screenshots, record audio conversations, capture keystrokes, and transmit files through a command and control server.

Attacks on the smart grid can occur at multiple levels including, transmission, distribution, and home networks. The attacks can include protocol-based attacks, routing attacks, intrusions, malware, and denial-of-service attacks. The attack vectors are varied including social engineering, random network scans, insider malicious activities, and physical destruction of the communication infrastructure.

1.2.2 Security Concerns

Smart grids consist of a network of sensors, monitors, devices, as well as computers for data collection and analysis. All of these are susceptible to cyber-attacks. Analysts have identified five major challenges faced by computerized security systems related to smart grids [7] including high volume of sensitive customer information, distributed control devices, lack of physical protection, weak industry standards, and a large number of stakeholders dependent on the grid. The concerns of smart grid security as with other typical systems are confidentiality, integrity, and availability. Confidentiality entails protecting both consumer and operations data; integrity is also required both at the consumer level for metering and billing and at the operational level to ensure stability of the grid; availability means that the power continues to be transmitted and received by customers, regardless of the status of the system.

Smart grid faces the same security challenges as any complex computer network and needs both perimeter defence and visibility into the network. The fundamental issue is that given massive size and interconnectivity in the entire network, worms and viruses can spread quickly. Also, given the distributed nature of the network, there are an enormous number of vulnerable targets. Additionally, SCADA systems are designed with inadequate security; for instance, Siemens still uses a hard-coded password for allowing access to control systems [8], which once compromised can lead to massive security breaches. Administrative passwords are often precoded and never changed from the original settings. There are several entry points into the networks, including infiltration through infected devices, network-based intrusion, compromised supply chain, and malicious insider.

There are several threats that the smart grid faces apart from dedicated attacks and intrusion by third parties [9–14], including privacy breach through data theft, electricity theft, disruption of services, physical damage to devices, denial of service, and market fraud. Hacking into smart meters, tapping wireless communication, or stealing the data from servers of the utility can provide fine-grained metering information of the users' consumption [9]. This information is necessary for the utility for billing, demand response, and load forecasting. The same information, however, can reveal the lifestyle of an individual. Each appliance has a unique electricity usage signature which can be extracted from the overall usage pattern indicating what the user is engaged in, i.e. working on a computer, watching television, taking shower, and cooking. Employers, marketers, insurance companies, as well as criminals can exploit this information for different purposes. Marketing companies could use this information for targeted marketing or introducing non-competitive pricing. Criminals can use this information to determine the daily routine or a family, i.e. when there is no one in the house or when someone is alone in the house for committing burglary or other crimes. Electricity theft can occur by altering the meter reading either by tampering with the meter or changing the information after breaking the encryption key [9].

1.2.3 Impact of Threats on Smart Grid

A small disruption (about 5 %) in communication can cause major latency issues leading to significant operational performance degradation [15]. Several metrics have been defined for communications in the smart grid including packet delivery ratio (# delivered/# expected), average end-to-end delay, and average packet hop (# of intermediate nodes), successful DR request ratio (# D-R requests delivered/# D-R requests issued) [15]. Limiting values of these metrics need to be defined and then guaranteed to ensure seamless performance of the grid. A key concern beyond communication latency issues is that data collected from sensors could be corrupted. There are mechanisms in place that can detect corruption of data based on other sensor values. An attacker can, however, manipulate data from enough sensors as to make data corruption unobservable [16]. Such attacks are not random but rather coordinated and not likely to be in sequence to avoid detection. For such attacks to succeed, the hacker would need knowledge of measurement detection and analysis techniques used at the control centre.

Smart grid relies extensively on wide area monitoring systems (WAMs), and the values from distributed sensors in the network are spatially analysed based on the GPS locations of the sensors [17]. GPS could be spoofed in measurement devices leading to wrong control decisions based on spoofed data, the outcome of which can be mild to severe based on the breadth of the attack. GPS can be spoofed by causing interference such that GPS receiver loses signal and then creating a false signal with a higher correlation peak that provides false information. False data can prevent fault signal from reaching the controller or provide a false location of the fault resulting in delay in power line repair and restoration. Voltage spikes can be camouflaged, and false voltage spikes can be generated leading to wrong corrective action by the controller causing instability in the grid. Coordinates of the disturbance can be falsified preventing triangulation and delaying the identification of fault location. Since message timing is crucial in smart grids, an attacker can use legitimate means to delay messages and cause denial of service or trigger faults. Attacker can flood the data stream with false data and severely degrade performance [18].

Summary
Ubiquitous communication is a necessary element of the smart grid, but it also provides hacker access to the grid components through the same network. There are security threats to the physical communication infrastructure as well as to the logical operation of the network based on conventional threats such as intrusion, denial of service, malware, and social engineering. Additionally, there are threats due to inadvertent errors, equipment failures, and natural disasters. There are several actors that pose a threat, including disgruntled employees, competitors, terrorists, nation states, and criminals. The entire smart grid is data driven where data is used for critical operations including, resource management, load forecasting, error correction, and fault isolation. Confidentiality, integrity, and availability are all very important in smart grid data security. There are numerous data-poisoning attacks that can destabilize the grid through unwarranted corrective actions or lack of necessary corrective actions, both of which can result in cascading failures. Lack of availability will result in a loss of visibility into the network that again is dangerous for the grid. In short, ensuring the security of the grid is critical to the success of the grid.

1.3 Ensuring Security in Smart Grids

The power grid is a very complex system that is geographically and logically distributed. The smart grid provides the communication infrastructure to connect the dispersed components and manage the grid by extensive data collection and analysis to get real-time operational intelligence. Such operational intelligence provides several benefits to the grid including improved load forecasts, peak load reduction through demand response, better utilization of renewable microenergy sources, and automated fault detection and isolation as well as correction in some cases. On the flip side, the communication infrastructure that permeates throughout the grid provides attackers access to the entire power grid. Consequently, it is imperative to have strong security in the smart grid.

The smart grid connects users, power plants, utilities, substations, and oversight bodies into the network with components, including protection relays and circuit breakers, SCADA systems, and household appliances. The smart power grid is distributed into three distinct segments, i.e. transmission, distribution, and HANs. In the traditional grid, the primary use of communication infrastructure in the distribution network was for monitoring substations. However, the communication network extends all the way to households and individual appliances with the smart grid. This also means that there is a much larger network to secure. Traditionally, the bulk distribution system has been the primary focus of cyber-security where the impact is the greatest. Failure on the distribution network has the possibility of triggering large-scale cascading failures. However, with the smart grid, attacks at the smart meter level can also have a large impact as attacks can spread through the network quickly leading to large catastrophic failures. There are several points of vulnerabilities in the grid [19] including the architecture, interoperability, communication protocols, interfaces, HANs, customer portals, and hardware.

A part of the problem today is the massive volumes of data being collected and analysed in distributed locations in the grid providing many targets for hackers for data manipulation attacks. A large part of the data volume comes from synchrophasors that provide the state information from the grid including voltages and currents required for ensuring grid stability. The data and software components of the infrastructure form a large chunk of the vulnerabilities that need to be addressed. Some of the conventional vulnerabilities come from validation checks in software including cross-site scripting, command injection, and buffer overflow [20]. Other vulnerabilities include poor management of access control, privileges, and permissions; lack of proper authentication; management of access credentials; and missing integrity checks. Other problems include poor configuration of systems, delayed patch management, lack of security audits, insufficient monitoring of logs, improper configuration of hardware and network devices, and finally lack of training of administrators in security practices.

There are a lot of legacy devices that were manufactured decades ago and do not have built-in cyber-security. During the transition period, when the devices are being gradually replaced, however, they form a large vulnerability. The past security paradigm in grid infrastructure was "security through obscurity", i.e. if the existence of a vulnerability is unknown, it will stay protected. We all know this is not true in the case of the Internet where networks are constantly being scanned for points of vulnerabilities. Also, as the software on SCADA systems get increasingly standardized, there is a chance of large-scale attacks through the network that can lead to large-scale failures and disruptions. Migration plan, thorough testing, and agile monitoring of the grid is necessary for ensuring that the legacy systems do not become a cyber-security liability for the smart grid.

1.3.1 Standards and Architectures

Standards are still evolving for smart grid appliances; consequently, security controls are being created differently for different devices preventing standardization in testing and evaluation. Several groups are actively working on creating standards, including Smart Grid Interoperability Panel (SGiP), Cyber Security Working Group (previously NIST Cyber Security Coordination task group—CSCTG), and Grid-Wise Architectural Council (GWAC). There are several requirements for security for smart grid that can be grouped into data security (access control, data authentication, storage, backup and recovery, and cryptographic protocols), security management (risk analysis, security policies, and training), and infrastructure security (system and device configuration, perimeter security, and personal key exchange) [21]. In addition, processes need to be developed to gain visibility into the network for extensive data logging and analysis. There is security need to be implemented through the communication infrastructure and systems, including SCADA (DNP3, GOOSE, IEC 61850, IEC 60870-5), WANs, land mobile radio (LMR), WLAN, and WiMax.

Most of the communication on the smart grid network would be encrypted with a need to use a public key infrastructure [22] in the grid. In addition, the communications infrastructure needs to be imbued with security incorporating, appropriate network topology design, secure routing protocols, secure message forwarding, end-to-end encryption, security broadcasting, and defence against denial of service (e.g. excess capacity, quick detection, and countermeasures). There also needs to be data packet authentication and bad data detection.

Numerous architectures have been provided for the smart grid communication networks. Reference [23] provides a 3-tiered architecture for the network including, HANs, neighbourhood area networks (NANs), and WANs and suggest use of a mesh area network that provides multiple redundant paths. Their architecture is primarily focused on preventing denial-of-service attacks and signal interruption. This architecture is agnostic to false-data injection attacks. Reference [19] suggests a layered approach to security for the smart grid that goes from technical execution at the lowest level to strategic direction at the top level, i.e. physical, network, host, data, application, business process, and enterprise organization.

1.3.2 Sensors and Devices

Individual smart meters need to be protected from tampering, data leakage, and intrusion. The hacker can gain access at the customer endpoint, crack wireless communications between the AMI meter and endpoint equipment, or crack wireless communications from the AMI meter to the local concentrator. Intrusion can allow access to the communication network of the utility through the endpoint. There have been several suggestions for their protection. One is to restrict transmission to only changes in power consumption; however, hackers can reconstruct the energy usage profiles from the power usage changes that are transmitted. There have been suggestions to include artificial spoofed packets into the data stream such that the energy usage looks normal rather than when an owner is not present. Spoofed packets can be randomly generated using Poisson distribution of power consumption or history templates [1]. At the transmission level, intrusion and buffer overflow type of attacks need to be detected. Most communication networks leave open a connection awaiting response to a SYN/ACK signal, sometimes as long as 75 s. An attacker can flood buffer with spoofed SYN requests creating congestion on the network. Bayesian statistical analysis can be used on the packet information to detect attack [24]. A fusion centre that uses transmitted data and library of previous data can also be used to determine whether malicious data are passed [25]. Each node will need to be analysed independently to protect against distributed attacks.

Most security models evaluate whether the current state of the system is valid by comparing it with a set of known security states. An exposure analysis graph can be used to identify users and data flows. Here, each node on graph has the following vertices: security mechanism, system privileges, information objects, and untrusted users; edges are directed paths to other nodes. This can be used to check for spoofing, tampering, repudiation, information disclosure or leakage, denial of service, and escalation of privileges [26]. Hierarchical Petri nets have been used to model multiple attacks [27]. Attack trees cannot track coordinated attacks, and multi-step Petri nets are limited to tracking three attackers. Hierarchical Petri nets are not limited to the number of attacks and can be used for multiple attacks including eavesdropping, interference or interruption of communication, unauthorized data access, service theft, and denial of service. The hierarchical model is built in teams such that local experts map the threat paths and outcomes in their areas, regional experts take local, mapping Petri net to network and create hierarchical structure and regional hierarchies combined into single overall Super Petri net using corresponding points.

Security of the physical state estimation is essential for the stability of the grid. Data collected from synchrophasors and other state estimation devices need to be analysed for corruption on malicious alteration. Security-oriented physical state estimation system [28] attempts to do that by exploiting the interrelation among the cyber- and physical components of the power grid. It utilizes information provided by alerts from bot host and network-based intrusion detection systems in its analysis. It uses file and memory check information from host-based IDS and permission issues, invalid signatures, and data packet inconsistency information from network-based intrusion detection systems to detect intrusion. It creates an attack graph template showing potential attack paths possible to be traversed by intruder and potential vulnerabilities. It works off base-case power flow solution, which defines how measurements should be correlated and checks for attacks using the template, and computes the probability that system has been compromised. Potentially compromised domains are noted and suspicious measurements identified. It then proceeds to suspicious measurements, attempts to estimate state while ignoring suspicious measurements, and if that is not possible waits for next interval to compute the state estimate. Reference [29] suggests different levels of protection of data based on criticality and providing the maximum security to a strategic subset of sensor measurements that influence the most system variables. Reference [2] suggests a comprehensive and integrated agent-based security platform with three layers of security, i.e. power, automation and control (monitors and control power grid processes), and cyber-security (handles access and data checking). Security agents located in meters, substations, and relay station command centers to handle protocol translation, security patch updates, pattern recognition, process flow, intrusion detection, data encryption, and access control. They propose using an anomaly-based detection system such that alarms are issued for activities outside of normal behaviour.

1.3.3 Network Security Threats

Several researchers have identified the various types of cyber-attacks that could threaten smart grid operations. The most exhaustive list was provided by [30], which includes eavesdropping, traffic analysis, interception of signals (electromagnetic and radio frequency), media scavenging, data interception and alteration, identity spoofing, bypassing controls, authorization violation, physical intrusion, man-in-the-middle, replay, malware, Trojans, trap doors, service spoofing, and resource exhaustion. A key threat to the grid is the potential for hackers to leverage the AMU for access to the bulk electric grid.

The main four that seem to be the focus of most research are eavesdropping, injecting false data via intercept/alter, service spoofing, and resource exhaustion. Some smart grid administrators do not even concern themselves with the spread of malware (like from viruses or Trojan horses) or the risk of a remote attacker assuming control of the system, believing that the firewall and other network protection on their computer system will be sufficient. However, many of these systems use HTTP and TCP/IP protocols, two systems that have documented vulnerabilities [31].

Eavesdropping is the situation where an outsider intruder listens or gathers data intended for the smart grid system. In this attack, the attacker, or eavesdropper, taps into the transmission signal between the data source (a home sensor, for instance) and the smart grid control centre. Eavesdropper can intercede between the time the data are encoded and the time it is decoded. That may slow down an eavesdropper, but some malicious attackers could have access to the common decoding algorithms, and with enough trial and error determine how to read the data.

Such successful decoding could then lead to the next type of attack: injecting false data. In this attack, the malicious intruder intercepts valid data and transmits false data to the control centre. Most control systems are decided to question or ignore data whose mean square difference from the normal or expected is too high [32]. Knowing this, though, an attacker can analyse data for a period of time, determine an acceptable range of values, and inject data that will be accepted by the control system [33]. The attacker can also serve as a "man-in-the-middle" and send fraudulent messages to either the customer or the system. Surprisingly, such an attack does not require much effort to cause an undetectable change in the system's operations. Experiments have shown that on an IEEE 300-bus system, it only took injecting bad data from ten different meters to cause an undetectable error that negatively affected most of the system control variables [32]. On most IEEE n-bus systems, it took as few as four strategically selected meters to cause such an error [32].

What is worse is that such attacks can be conducted from a variety of sources. An individual meter can be attacked, causing it to transmit corrupted data or causing it to stop transmitting entirely. A substation, which collects data and monitors distribution for a particular region could also be subject to attack. A substation attack can involve blocking data from certain sources, injecting false command codes, or misrepresenting the power flow into or out of that substation [2]. Even the control centre itself is not immune. If an intruder can gain access, the SCADA could be flooded with bad data, a communication link with a substation (or series of substations) can be broken, command codes could be altered, and consumer price rates can be changed [2].

Response and recovery engine [34] employs 2-player adversarial Stackelberg stochastic game theory along with attack–response trees that create Markov decision trees for intrusion prevention, detection, and response. There are three main types of intrusion response systems, i.e. lookup tables with predefined mappings, which are neither scalable nor flexible, and heuristic based, which could become predictable to the intruder and selection models. They suggest an engine with a state space large enough for decision analyst to be able to create attack–response trees that uses a multi-step process for the response, i.e. determine what areas have been attacked, identify appropriate attack–response trees for the attacked areas, create responses by collapsing response sequences into Markov decision processes (resolve uncertainties using Bayes binary classification), and determine best action to take based on chosen responses and system criteria. The process can be repeated for each new attack.

Several security systems detect an error or attack and trigger an alarm, but are not designed to adaptively fix and prevent future attacks. Reference [35] focuses on preventing future attacks and suggest an anomaly security system that uses past normal data as well as data with intrusion to update anomaly classification information. It also suggests using instruction set randomization to prevent code-Injection attacks and a transformational key such that injected code does not mesh with the rest of the code. This can also prevent man-in-the-middle and denial-of-service attacks. Anomaly classifications help identify bad data injection which is not perfect and will lead to false positives. They suggest using false-positive correction as bad data for an attack. Their model suggests a 3-step process, i.e. filtering to trap any suspicious activity, classification to evaluate malicious behaviour and supervision to provide feedback to proxy/agent, and remediation to prevent future attacks.

Summary

Cyber-security in the smart grid is required at the perimeter as well as internal to the network. The standard perimeter defence would include firewalls, intrusion detection systems, and secure architecture, while the internal defence would include integrity checks, network monitoring, and log analysis. In addition, it is necessary to institute a key exchange mechanism along with protocols for end-to-end encryption of data. It is also necessary to institute robustness to false-data injection (FDI) and denial-of-service attacks by creating redundant channels and fall-back positions for state estimation and load forecasting.

1.4 Mitigating Cyber-Physical Threats

One of the key security unknowns is how vulnerabilities can be exploited in the cyber-physical domain, i.e. can a cyber-vulnerability lead to an attack on the physical infrastructure or vice versa can a physical vulnerability expose an attack on the cyber-infrastructure. It is anticipated that the SCADA systems be targets of multifarious attacks from several actors including foreign governments, terrorists, and competitors. SCADA systems are typically engaged in data collection, analysis, control, and visualization. Such systems would be used not only for the traditional operation of the power grid but also for smart grid-specific applications including enabling microgeneration, automated recovery from faults, enabling electricity market functions (price signalling, energy trading), and demand response (DR). Its enhanced capability would also make them ripe targets for cyber-attacks. The typical modus operandi of a cyber-attack would involve the following: (1) gaining access to the SCADA network either through a corporate network, VPN connection, or a remote site connection, (2) probing the SCADA network to discover the appliances, data storage, and vulnerabilities and deduce the SCADA processes, (3) attacking and controlling the SCADA system by gaining root privileges, getting access to the data, and launching control commands.

A typical cyber-physical system attack would involve four steps: (1) identifying weaknesses in the cyber-infrastructure; (2) intruding into the system and gaining privileges; (3) understand and gaining control of the control system; and (4) using the control system to launch physical attacks. One of the key concerns is data manipulation at destabilization of the grid as well as denial of service. For instance, in case synchrophaser data are manipulated through FDI, the grid could be made to oscillate and eventually go down. The demand can also be manipulated forcing a demand–response from the utility company effectively denying the availability of power for some consumers [36].

There are several potential attacks that can be launched against SCADA systems including false-data injection, replay attack (forging time stamps), denial of service, and sensor spoofing. For instance, the smart grid will have automatic detection of anomalies—if a false anomaly is injected into the grid, it could lead to dispatch of crews in unneeded areas. Most importantly, attacks in a substation can include missing or corrupted sensor data as well as false-command injection and delay in data transmission. Such attacks can cause circuit breakers to open at the wrong times, system run exceeding limits, system outage, false alarms, damage to equipment, and injuries/deaths or operators and end-users [36]. There can also be attacks that corrupt data going from transducers in the field causing circuit breakers to trip and leading to outage. Attacks could also include tampering metering data that can lead to false implication of users resulting in penalties including fines and termination of connection. Often, substations are controlled from control centres, and any falsification of communication data between the two can lead to system outages, false alarms, incorrect procedures, system outages, and physical injuries.

Many of the software for SCADA systems were developed decades ago without security considerations, making SCADA systems highly vulnerable to software exploits. A lot of software does not have adequate authentication and access control mechanisms, making access to hackers easier. Due to the large number of vendors and devices, it is difficult to test all the devices and software ahead of time. Over the last two decades, we are seeing more homogeneity in SCADA system software that allows for better testing and validation of software for security compliance. This homogeneity, however, is a mixed blessing—having obscure operating systems and devices makes generic attacks harder; however, it makes targeted attacks by dedicated adversaries easier [37]. Since late nineties, there has been a strong focus on standardization of SCADA systems leading to greater homogeneity which makes them targets for mass non-specific attacks and probes. Coordinated large-scale attacks will be facilitated by the homogeneity in the network that can overcome the resilience of the network and cause large-scale failures in the grid.

1.4.1 Risks at Cyber-Physical Interface

The risks at the cyber-physical interface follow the logical divisions of the smart grid infrastructure, i.e. generation, transmission, and distribution [38]. At the generation level, the risks occur at the level of automatic voltage regulation, governor control, and automatic generation control.

1.4.1.1 Generation

Power carried in alternating current networks is typically comprised of real power and reactive power. The real power is used for doing work, while reactive power is used for maintaining voltage stability. By controlling the production, absorption, and flow of reactive power, voltage can be maintained within acceptable limits, while transmission losses are minimized. Generator exciter control is used to control the amount of reactive power being absorbed or injected into the systems. The control module communicates with the plant via Ethernet, and by comparing the generator voltage output and voltage set points, it alters the current flow through the exciter to maintain stable voltage. Similarly, governor control is used to control the frequency of the rotor by altering the power output from the generator. Again an Ethernet connection is used to measure the rotor speed and provide feedback to the governor control for altering the power output. Both of these control systems are local without requiring remote telemetry; however, there are vulnerabilities associated with malware that can be inserted locally through USB or by compromising the local area network. Altering the set points or injecting false data on the output readings can lead to instability of the generator.

Another area of concern is the automatic generator control wherein output from multiple generators is adjusted for changes in the load. The output from the generators must match the anticipated load on the grid very closely or else consumers would experience voltage sags and spikes which are both bad for operation of electric and electronic equipment. The balance can be estimated by measuring system frequency. Increasing frequency means more power is being generated than used, and vice versa decreasing frequency means more load on the system than the generators are producing. Automatic generator control increases or decreases load across multiple generators based on prior protocols. An attack on the automatic generator control can result in significant operational damage through instability in the grid. Since multiple generators are involved, there is obvious need for remote telemetry to gather load data and provide feedback to the generators. This increases the vulnerabilities in the network that can include disruption of telemetry, false-data injection, intrusion, and denial of service.

1.4.1.2 Transmission

At the transmission level, there are two applications that are critical, i.e. VAR compensation and state estimation. VAR compensation is done using fast-acting devices for providing reactive power on high-voltage transmission lines for impedance matching. If the grid's reactive load is leading, VARs are consumed to lower the voltage, and if the reactive load is lagging, the capacitor banks are switched on to increase the voltage. The modern VAR compensation devices are thyristor controlled that can operate autonomously. There is a network of such devices that need to communicate with each other to determine the operating point. A denial-of-service attack on the network could result in an inability to communicate impacting the dynamic control capabilities causing degradation of power quality or disruption of power due to voltage sags and surges triggering shutdown of critical devices. There could also be timing-based attacks that disrupt the synchronization of the devices, which is critical for operation of the network. Finally, there could be data injection attacks that send incorrect operational data that may result in incorrect VAR compensation impacting the synchronization.

To improve the situational awareness of the electric grid and to maintain the stability of the grid, the state of the grid needs to be monitored. The new smart grid will also be retrofitted with synchrophasors. These devices measure the characteristics of the electrical current travelling at different points on the grid at short time intervals (typically 30 measurements per second). They typically use a common time source typically based on GPS to allow for time synchronization across the entire grid. This data will facilitate a number of applications while enhancing others, such as real-time monitoring of the system, state estimation, disturbance monitoring, instability prediction, and wide area protection and control. The characteristics of the data generated by synchrophasors make them particularly vulnerable to cyber-attacks. They play a critical role in maintenance and control and power generation and distribution, making them attractive targets for malicious actors for disrupting the power grid. Synchrophasor data are collected at geographically diverse locations and are usually routed to data concentrators in central locations using public Internet, making it susceptible to several attacks including FDI, disruption of communication, and corrupting the analysis. One of the attacks is based on data analysis where a hacker has access to partial data which can be analysed by a hacker to predict behaviour of the grid and then use the information to attack the grid. There are obvious ways in which the data can be protected including data obfuscation, anonymization, and encryption.

1.4.1.3 Distribution

The distribution system carries lower-voltage power across distribution lines to the customers. This system will have several applications that will have intelligence built into it. The most visible applications are the Advanced Metering Infrastructure (AMI) and DR. The AMI will allow for increased reliability, incorporating renewable integration from microenergy sources, and provide visibility into the usage at the customer end down to the appliance level. Smart meters will provide utilities with load control switching (LCS) ability to turn off appliances during peak hours to better balance the load. The smart meters pose strong vulnerability at individual consumer level whereby services could be disabled or enabled by hackers at will if they were to breach the security of the smart meter. The second major application is billing application for which the smart meters will read usage data, validate it, and create electricity bills. In addition, the meters will be used to establish and terminate services as well as restrict services for non-payment of dues.

A second key application at the distribution level is the self-healing elements of the grid where automatic reclosers are used to clear momentary faults. Faults that cannot be autocorrected can be detected through sensors placed in the distribution network. Data injection attacks can be used to show spurious attacks that will lead to unproductive dispatch of resources. At the same time, a denial of service on the network can prevent crews from reaching a site of an actual disruption.

1.4.2 Mitigating Cyber-Physical Threats

The fundamental problem with the smart grid is its geographic expanse across a vast area with several soft targets that are vulnerable to attacks. Physically defending the entire grid is a daunting task; consequently, building resilience in the infrastructure is a critical mitigation strategy, including self-recovery, redundancy in power distribution and communication, excess capacity in communication, power conduits, and physical power hardware. The second critical issue is to ensure that the critical control systems have a system of alerts that quickly provides alerts when the device is operating at a dangerous level. Thirdly, we need to deploy manipulation detection algorithms on a case-by-case basis of different algorithms to minimize the impact of data poisoning. Perfect security is unachievable; however, the goal is to minimize the risks such that malicious activity can be detected quickly, catastrophic situations can be avoided, and recovery from attacks and anomalies can be swift. We recommend a risk analysis approach to understand the high-level exposure to the smart grid and mitigate the threats that it faces. Cyber-physical threats require creation of detailed attack trees to understand the cyber-physical interactions in the grid.

Summary

There is considerable danger to the smart grids associated with the cyber-physical threats. We have risks at each level of the grid, including generation, transmission, distribution, and home networks. The scope of damage varies from catastrophic to minor based on the attack vector and where it is launched. The increasing homogeneity and connectedness in the grid provides a fertile ground for launching large-scale attacks that can have serious repercussions on the operations of the grid including large-scale blackouts. Our policy has to be quick detection and containment for defending against zero-day attack vectors, and for the run of the mill cyber-attacks, we need to develop redundancy and resilience into the grid to prevent catastrophic failures.

1.5 Mitigating Smart Meter Threats

1.5.1 Threats and Vulnerabilities in Meter Infrastructure

As the key components in smart grid infrastructure, smart meters accommodate the most valuable data (e.g. meter readings) for improving the performance of power grid and changing the lives of electricity consumers. For instance, meter readings are required to support many smart grid applications and services, including automatic meter reading, billing, dynamic pricing, and detection of impending blackouts and energy thefts, which can bring great convenience to both utilities and energy consumers. However, the massive amount of data collected from smart meters should be carefully protected against misuse. It is desirable to incorporate security mechanisms into the design and implementation of smart meter infrastructure so as to increase robustness and resilience for the system and gain energy consumers' trust.

Skopik et al. [39] analysed the security threats and vulnerabilities in smart meter infrastructure detailed in three tiers: smart meters, utility, and Web application. The first-tier smart meter vulnerabilities are categorized as the attacks to the smart meters (devices) itself, such as manipulating the hardware and the firmware, and exploiting limitation design and implementation. The corresponding countermeasures and defence mechanisms for such attack include authentication and strong encryption of communication, secure key management, securing the firmware, and secure source code development. The second-tier vulnerabilities occur at the utility, which suffers the potential attacks such as near-me area network (NAN) sniffing, own or foreign meter emulation, large-scale meter takeover, and concentrator nodes (s) attacking. Secure system design, secure operation, and secure service evolution can be utilized to tackle such security concerns at the utility [3]. The last tier, Web application vulnerabilities can be exploited by attackers by compromising the security in smart metering data management and value-added services, such as automatic billing, as well as the privacy in smart grid [40].

In smart grid, the AMI accommodates two-way communication between the smart meters and utility and enables remote control and monitoring for both energy service providers and consumers. Rahman et al. [41] investigated the non-invasive threats and the vulnerabilities in such infrastructure, such as lack of authentication, slave meter data tampering, slave meter unauthorized disconnection, insecure protocol implementation, and firmware upgrade vulnerabilities. More specifically, the non-malicious threats involve reachability and integrity threats, and availability threats (e.g. improper scheduling of data delivery between meters and collectors leads to buffer overflow and data loss in the collector side). The malicious threats could be typical cyber-threats on AMI such as DoS, link flooding, and wireless link jamming. For instance, a large number of compromised collectors can launch a distributed DoS attack to a headend. In this scenario, it is infeasible to resolve the cyber-threats from the compromised collectors. Indeed, the heterogeneity of interdependent hardware configurations (each operating with various security parameters) would lead to both malicious and non-malicious attacks [41]. Rahman et al. have proposed the detection methods in an automated security analysis tool for AMI—SmartAnalyzer. It provides the following functionalities [41]:

- Extensible global model abstraction capable of representing millions of AMI device configurations.
- Formal modelling and encoding of various invariant and user-driven constraints into SMT logics.
- Verifying the satisfaction of the constraints with AMI configuration using an SMT solver [26, 30].
- Identifying potential security threats from the constraint violations and providing remediation plans for security hardening by analysing the verification results.

With AMI, meters are not read manually anymore, but digitally instead. The digital usage rates transmitted from site to site would leave loopholes and security vulnerabilities for malicious attackers and energy theft. Xiao et al. [42] have also identified three classes of attacks based on when and where the data for the amount of service are manipulated: (1) while the data are recoded, (2) while the data are at rest in the meter, and (3) as the data are in flight across the network. They discussed the possible way to resolve these types of attacks by installing a redundant meter at the energy provider end. However, the above solution is impractical because of the huge number of "inspector" meters required for all the end-users (each user needs one). Alternatively, Xiao et al. [42] proposed a model, in which N number of end-users' meters are monitored by a "head inspector". The head inspector utilizes a series of algorithms to collect heuristic usage information based on an adaptive-tree-based inspection scheme. The inspection strategy in response to anomalous readings can be adjusted to pinpoint the meters where fault or security compromise occurs. This strategy is effective to address the aforementioned classes while maintaining a low cost compared to monitoring each meter directly.

1.5.2 Security Breach on Smart Meter

Similar to other contexts of security, Gering [43] discussed the confidentiality, integrity, and availability of smart meters. Specifically, "confidentiality" ensures that sensitive data are not exposed to the unauthorized person or system and the information disclosure should be limited. "Integrity" ensures that actions can be traced to initiators, which helps to protect against deception. "Availability" ensures that data, commands, and communications are accessible and usable when desired. To guarantee each of them, for instance, Gering [43] stated that encryption techniques can be used to ensure confidentiality using techniques such as triple data encryption algorithms, advanced encryption standards, elliptical curve cryptography, and RSA public key cryptography.

Some examples of breaching different aspects of security are given below:

- To hack into a smart meter, David Baker (the director of service at IOActive, a Seattle-based research company) described a possible way to pass through the smart meter's wireless networking device. A software radio, which can be programmed to emulate a variety of communications devices, can be used to listen wireless communications with the network and deduce how to communicate with the meters over time. Besides this, he also discussed another method —attacking the hardware. An attacker could steal a meter from the side of a house and reverse-engineer it. However, this method requires a good knowledge of integrated circuits for reengineering the meter, which is inexpensive [8].
- An independent security researcher specialized in wireless sensor networks, Goodspeed (an independent security researcher) told another story about smart meter hacking [8]. If the meter has not been built with rigorous security features at the physical level, a hacker can insert a needle into each side of the device's memory chip. It is indeed a probe to intercept the electrical signals in the memory chip. Then, the hacker can readily obtain more information from the device by analysing such signals. Even if some security features have been integrated into the meter, it may be possible to extract the information using some customized tools.

- Besides the inevitable smart meter hacking activities, a massive network virus or worm can also attack utilities. In this case, utilities can implement granular security architectures to protect their smart grid system [44]. The unique standard-based hardware and software security should be embedded into the network node and device. Such security modules could help prevent device penetration attacks (in the form of worms or viruses) from spreading throughout the network. The embedded device-level security ensures that a hacked or compromised device can be quickly identified and isolated before spreading or causing greater damage. Including the above case, utilities leverage the best efforts they made and millions of dollars of investments on smart grid/meter security to the latest security technologies in other contexts.

1.6 Mitigating Data Manipulation Threats

1.6.1 Introduction

Cyber-security in critical infrastructure and particularly the smart grid has received significant research interest [45–47, 30, 43, 48, 48]. In modern smart grid, Supervisory Control and Data Acquisition (SCADA) software and hardware component is generally implemented to supervise, control, optimize, and manage power generation and transmission. The SCADA system integrates new components (e.g. smart meters), networks, sensors (e.g. phasor measurement units or PMUs), and control devices. More intelligently, the future smart grid infrastructure will accommodate renewable energy resources, electric vehicles loads, and storage, among others [50] by making the components intensively interconnected. However, new vulnerabilities may arise along with the convenience brought by new features in smart grid. So far, hackers begin to penetrate the control network and administrative devices in the US electric grids via Internet [51]. In August 2010, a computer worm targeted the SCADA system, infected thousands of computers, and tried to compromise the critical infrastructure [52].

As a centralized control centre which conducts controlling and monitoring activities for the power grid, SCADA system receives and stores various real-time meter measurements, including bus voltage, bus real and reactive power injections, and branch reactive power flows in every subsystem of a power grid. State estimation plays a key role in controlling- and monitoring-based energy management in SCADA system [14, 53], which optimally estimates the state of the grid by analysing data such as system parameters, power meters, and voltage sensors. More specifically, such function estimates unknown system variables using the meter measurements data in the electric grid. Results of the state estimation will be generated to maintain system in normal state, to optimize the power flow such as increasing the yield of an electric generator, to balance supply and demand load, and/or to ensure reliable operations such as detecting faults in the system [54].

A malicious adversary may aim at altering the data (e.g. meter readings) transmitted to the control centre. Thus, such violation of data integrity will result in great threat to the entire smart grid system since the decisions of energy management created in system estimation might be significantly deflected by this kind of malicious behaviour, namely FDI attack. Essentially, FDI attacks maliciously modify the data generated in smart grid (transmitted to and stored in SCADA system) and may potentially trigger two negative impacts [36]:

- If the data are modified in a way that is not detectable as false by state estimation, the observable state of the system will be wrong and may lead to actions by the grid operator where security concern may arise in the system.
- The malicious intent may not be able to hide the attack. Even though the attack is detected, part of the system may become unobservable, which means that the state estimator cannot estimate state values such as voltage magnitudes and voltage, and the transmission grid would be vulnerable to a local physical attack. By the time the consequences of the physical attack have propagated into the rest of the system where the state is observable, it may already be too late to avoid an outage of a larger part of the system.
- Data manipulation threats and FDI attacks would explicitly or implicitly lead to significant errors by compromising the meter readings in state estimation (optimal estimation of the power system state using data from power meter voltage sensors and system parameters [50]) or other smart grid components. Roughly speaking, FDI attacks can be categorized into the following two types [54]
- Observable/non-stealth attack: naive false-data integrity detection algorithms can easily detect such attacks since only meter measurement data have been changed. Difference between the compromised data and the physical information could be used to detect and report such kind of attack by the control centre.
- Unobservable/stealth attack (the compromised meter readings are consistent with the physical power flow constraints) will bypass many false-data integrity detection algorithms.

In this chapter, we summarize the potential data manipulation threats of FDI attacks in smart grid (particularly the unobservable attacks). The state-of-the-art defence mechanism or countermeasures are proposed to detect and tackle the threats as well as system vulnerabilities.

1.6.2 Resolving Data Integrity Violation in State Estimation

Compromising meters at the control centre and introducing malicious measurement has been discovered as an attacking technique for adversaries recently [55]. For instance, an online video tutorial shows people how to manipulate electric meters to cut the electricity bills (https://www.youtube.com/watch?v=wa13_l-qjBE). Following the same instructions, it is possible that the attackers target the meters at the smart grid control centre and inject bad measurements. If the outcome of state estimation is altered by the adversaries with such injected bad measurement, severe incidents such as power outage of large geographic areas may occur.

Some researchers have developed techniques to identify and tackle the observable malicious measurement injection [56, 57], where most of the techniques were targeted at arbitrary, interacting/correlated malicious measurements. More recently, more practical and advanced problems on attacking smart grid state estimation are investigated. For example, Liu et al. [32, 55] discovered that if prior knowledge such as the configuration of the power system is known to the adversaries, malicious measurement could bypass the regular detection and identification techniques proposed for observable attacks. Observable malicious measurement attacks could be easily detected because "the difference between the observable measurement and the estimated measurement becomes significant" [56]. Liu et al. [32, 55] studied a new class of threats to state estimation, namely FDI, assuming that the adversary can take advantage of the power grid configuration from the perspective of the attackers. They showed that the attacker can inject malicious measurements that can bypass the bad measurement detection on observable attacks and focused on two realistic attacking scenarios:

1. The attacker has limited access to some specific meters
2. The attacker is limited in the resources required for compromising meters. Specifically, two attacking goals are considered in [32, 55], which are random FDI attacks (injecting a random error to the result of state estimation) and targeted FDI attacks (injecting an arbitrary error to the result of state estimation).

Note that in the above work, the attacker is assumed to know the target power grid configuration and the meters are manipulated before they are used for state estimation, possibly as an insider or ex-insider. Although strong requirements are posed in the scenarios, the electrical engineers and security personnel should be aware of the threat which would lead to catastrophic impacts as well.

Rather than assuming that an adversary possesses complete knowledge on the power grid topology and transmission line admittances [32, 55], Rahman and Mohsenian-Rad [58] investigated a more practical scenario in which the attack has limited information with respect to the power network topology or admittance for some transmission lines. They disclosed that it is possible to compromise state estimation with only incomplete information against smart power grids. A more realistic FDI attack was introduced in [58], where various grid parameters and attributes such as the position of circuit breaker switches and transformer tap changers are unknown to the potential adversaries, and the adversaries also have limited access to most of the grid facilities. Covertly compromising the readings of multiple power grid sensors and PMUs in order to mislead the operation and control centres was identified as the major threat against smart power grids in [58], though adversaries only have incomplete information. Moreover, two types of FDI attack were introduced in [58], which are perfect attacks (the attacker has complete knowledge of the admittance for all lines on at least one cut on the grid topology) and imperfect attacks (the above information is not available). Rahman and

Mohsenian-Rad also showed that it is possible to construct a probability distribution function for unknown admittance to design an imperfect attack and simulated the result with a novel vulnerability measure.

For "unobservable attack", Kosut et al. [16, 59] distinguished two primary regimes in which malicious unobservable data attacks occur, by whether the attackers have controlled sufficient meters to commit the unobservable attack. They discovered that two regimes have completely different behaviour to corrupt state estimation [16].

1. Strong attack regime: adversaries are able to access a sufficient number of meters to commit an unobservable attack. Attacks cannot be detected by the control centre, even if there is no measurement error.
2. Weak attack regime: adversaries do not have access to a sufficient number of meters; the attacks can be detected, though imperfectly due to measurement errors.

Kosut et al. studied the behaviour and presented the results of both regimes in [16, 59]. Also, from the perspective of the attacker, Kosut et al. [59] investigated that how vulnerable a power system is to the unobservable attack. More specifically, they explored the smallest number of compromised meters required to perform the unobservable attack and presented an efficient algorithm to find the small sets of meters required for triggering such attacks based on the purely topological conditions for observability (graph-theoretic approach). They also examined the worst malicious data attacks in the regime that the adversary cannot perform an unobservable attack. In [16], another relevant problem from the perspective of attackers was studied is examining the trade-off between maximizing the estimation error at the control centre and minimizing the detection probability. Besides the graph-theoretic approach presented in [59], detection mechanisms and counter-measures are proposed for the weak attack regime in [16]. Specifically, since the adversary can choose where to attack the network and design arbitrary injected data, hypothesis test cannot be used for formulating the malicious data detection problem. Instead, a detector based on the generalized likelihood ratio test was proposed, which is known to perform well in practice. If the detector has sufficient data samples, the performance is close to optimal. However, solving a combinatorial optimization problem is desirable for the detector; thus, if the number of corrupted meters is large, it is difficult to implement the detector due to efficiency. To tackle this issue, another detector is studied—using a convex regularization of the convexity of the optimization problem based on L1 norm minimization.

Giani et al. [50] tackled another specific unobservable attack problem for smart grid state estimation—unobservable low-sparsity cyber-attacks, which require coordination of a small number of (≤ 5) meters. Since cyber-attacks of large number of meters in control centre tends to be improbable (for the reason that high degree of temporal coordination across geographically separated attack points is required for unobservable attack), they proposed an efficient algorithm to find all unobservable attacks involving the compromise of exactly two power injection meters and an arbitrary number of power meters on lines. The algorithm requires $O(n2m)$ flops for a power system with n buses and m line meters. If all lines are metered, there exist canonical forms that characterize all 3, 4, and 5 sparse unobservable attacks. These can be quickly detected with $O(n2)$ flops using standard graph algorithms. Known-secure phase measurement units (PMUs) can be used as countermeasures against an arbitrary collection of cyber-attacks.

In some occasions, simultaneous attacks may occur on multiple meters of electric grids to manipulate state estimation. To formally formulate this type of data injection attacking problem, Kim and Poor [60] presented a unified formulation for the problem of constructing attacking vectors under an optimization framework by considering constraints on the measurements and limited resources of the attacker. Linearized measurement models were given against the attacks of manipulating system state estimators. They also showed that the proposed approach significantly outperforms the prior work.

1.6.3 Resolving Other Data Manipulation Threats

1.6.3.1 Topology

As an important input to smart grid operations, topology of smart grid includes state estimation, real-time pricing, and real-time dispatch [33]. Adversaries could partially manipulate the grid operations by perturbing the topology information of smart grid. Although topology information involves the data for power grid state estimation, topology attack may have different behaviour and targets from the FDI attack committed for state estimation. For example, an adversary may mask a connected line as disconnected or vice versa so that the control centre makes improper decisions in contingency analysis, optimal dispatch, or load shedding [33]. Moreover, since topology information can be used for computing real-time locational marginal price, adversaries may modify the topology estimate to maximize the adversaries' gain. Thus, besides state estimation, topology of smart grid is vulnerable to malicious data injection attacks.

Kim and Long [33] focused on the man-in-the-middle attacks applied to topology of smart grid, where the adversary intercepts network data (e.g. breaker and switch states) and meter data from remote terminal units, partially modifies them, and forwards the maliciously modified data to the control centre. Similar to "observable attack", if not both network data and meter data are altered in the attack, modern power systems equipped with bad data test could discover such inconsistency. Therefore, the adversary is assumed to successfully bypass the bad data test by modifying both network and meter data (consistent with the "target" topology) with known global information about system state. Similar to the state estimation attack, the feasibility condition for undetectable attacks was given along with the low detection probabilities in [33].

1.6.3.2 Load

Adversaries may commit cyber-attacks to electricity generation, distribution/control, and consumption in smart power grids. Compromising state estimation (as summarized above) indeed attacks the electricity distribution/control. Mohsenian-Rad and Leon-Garcia [9] investigated a typical data manipulation threat in the consumption sector—the load might be modified by adversaries. More specifically, with the development of demand-side management and the growth of Information Technology integrated into consumption, altering the load at specific grid locations through the Internet and by distributed software intruding agents has been identified as a new class of cyber-intrusions. Such data manipulation threat may involve abruptly increasing the load at the most crucial locations in the grid and then cause circuit overflow, or other malfunctioning that can immediately bring down the grid, or significant damage to the power transmission and user equipment.

Specifically, such attack called "Internet-based load-altering attack" is defined in [61] as follows. An Internet-based load-altering attack is an attempt to control and change (usually increase) certain load types that are accessible through the Internet in order to damage the grid through circuit overflow or disturbing the balance between power supply and demand. Notice that three types of loads are accessible through the Internet and can be the target of load-altering attacks [61]:

1. Data centres and computation load: a data centre's power load is highly elastic and relies on the data centre's computation load. The energy consumption of data centre can be doubled when computer servers are busy, compared to when the computer servers are idle. Thus, data centre can be the appropriate target of Internet-based load-altering attack.
2. Direct load control: with Internet-based load-altering attack, the adversaries may compromise the command signals to seize the operation of the residential and industrial load which are supposed to be controlled by direct load control programs (one of the most common demand-side management programs used for minimizing peak demand, improving system operation, or maximizing quality of service).
3. Indirect load control: in smart grid, indirect load control allows customers to control their loads independently in terms of the price signals sent by utilities, e.g. through the Internet. Given the price information and based on the energy consumption for each household appliance, the decisions can be made by minimizing the cost of energy, minimizing the finishing time for the operation of appliances, or achieving a desired trade-off between cost and timing. Since the price information is obtained through the Internet, load-altering attacks can inject false-price data into the automated residential load control. Major changes of the load profile can be caused by modifying the energy consumption program in thousands of households.

Essentially, Mohsenian-Rad and Leon-Garcia [61] overviewed a collection of defence mechanisms which can facilitate blocking the Internet-based load-altering attacks or mitigating the damage caused by such attacks. The defence mechanisms range from protecting the command and price signals in direct and indirect load control to load shedding, attack detection, protecting smart meters, and load relocating. To reduce cost for applying defence mechanisms, the authors proposed a cost-efficient load-protecting strategy to minimize the cost of load protection while preventing from overloading the grid.

Summary
In summary, data manipulation threats may exist in most data-intensive components in smart grid infrastructure. How to detect the FDI attacks (both observable and unobservable), and eliminate or mitigate the vulnerabilities in smart grid have attracted considerable interest in smart grid research. As a primary data manipulation threat to smart grid infrastructure, the FDI attackers intend to mislead the decision-making of smart grid by hacking the readings of multiple sensors and PMUs. FDI can be executed to the smart grid components and devices in which data are generated, transmitted, received, and stored. For instance, state estimation requires data analysis received from meters; thus, data collected from the meters will be the target of potential FDI attacks, vulnerable to data manipulation threats.

In this chapter, we illustrated the behaviour and characteristics of the data manipulation threats and attacks according to their targets such as state estimation [16, 59], topology information [33], and load at the energy consumption side [61] and briefly introduce the defence mechanisms and countermeasures proposed in the literature.

1.7 Mitigating Privacy Threats

1.7.1 Introduction

Today, enormous amount of data/information are ubiquitously collected by commercial companies, organizations, or governments for analysis, which facilitates the development of services and applications in many industries. In practice, it is often necessary for the data owners to share their data to other parties for functioning the corresponding services and applications, or deriving more comprehensive and precise knowledge. However, explicitly sharing data would incur significant privacy risks to the individuals or organizations. Some serious privacy-leaking incidents happened recently; for example, AOL Inc. published their customers' 3-month Web search history in 2006 for research purpose. Although the IDs have been removed before data publication, many AOL users were still identified from their search information by the adversaries, and then, much of their private information and personal behaviour were exposed to the public. Also in 2006, Netflix Inc. published their customers' movie rating information to accommodate an open competition for the best collaborative filtering algorithm of predicting users' movie ratings. In 2007, two researchers from the University of Texas identified individual users from the Netflix movie rating data by linking the datasets to some other sources such as Internet Movie Database.

Such incidents exist almost everywhere, such as healthcare systems, location-based services, and DNA applications. Smart grid has a similar story as above on privacy threats. More specifically, implementing "smart" in modern grid systems requires information disclosure across different parties, many of which are untrusted in general. For example, utilities need to monitor electricity usage and load and determine bills; electricity usage advisory companies need to access the metering information to promote energy conservation and awareness; marketers access the profile of the customers for targeted advertisements; law enforcement officers access smart grid data for criminal investigation [37]. All of these data access may comprise consumers' privacy in smart grid system. Precisely speaking, utility usually collects the fine-grained energy usage (perhaps at the appliance level) from their customers, where the households' personal behaviour could be learnt from the status of appliances [41, 23].

On the one hand, consumers wish to save energy and their money with smart grid applications. However, on the other hand, they worry about the private information leakage since an intelligent monitoring device transmits their live usage to utility every 15 min with smart metering service [37]. Besides the personal behaviour patterns learnt by strangers, metering information disclosure may also make them vulnerable to annoying advertisements, thieves, or even robbers (e.g. criminals can identify the best times for a burglary or to identify high-priced appliances to steal [37]). A report released in 2010 by the consulting company Accenture states that one-third out of more than 9000 consumers from 17 different countries are not comfortable to use energy management programs provided by smart grid (e.g. smart metering) if their personal consumption information could be easily accessed by utilities [37]. Therefore, it is desirable to design smart grid services and applications without compromising individual customers' privacy and organizations' proprietary information. In this chapter, we investigate the privacy issues in smart grid infrastructure by illustrating the privacy threats, privacy laws, and state-of-the-art schemes related to smart grid.

1.7.2 Privacy Threats in Smart Grid Infrastructure

Personally identifiable information (PII) is the information that can be used on its own or with other information to identify or locate an individual person. PII can be one's name, contact and biographical information, individual preferences, trans-actional history, activities, or any information derived from the above [62]. In the context of smart grid [47, 62], at the customers' end, the linkage of any PII and the energy consumption could be utilized to identify individuals. Many customers' activities and end-user components may disclose their personal information to utilities or other untrusted parties, such as smart meters, smart appliances, dynamic pricing, load management, and consumer access to energy-related information [62]. For example, smart appliances communicate frequently with the grid to share the real-time energy usage information as well as the status of the appliance; dynamic pricing provides the current or future pricing information to customers and enable them to modify their demand at different time (e.g. time-of-use pricing, critical peak pricing, real-time pricing)—the preferences and response could indicate the personal behaviour and help identify customers.

A senior consultant with Cutter Consortium's Business Technologies Strategies practice and privacy professor, Rebecca Herold identified and discussed the data privacy concerns in the smart grid in the NIST SmartGrid privacy group report [63]. The privacy concerns w.r.t. PII are summarized as below:

- Identity Theft: the combination of PII may be misused to impersonate a utility or consumers, resulting in potentially severe threats. Attackers can masquerade them to forge negative credit reports, behave fraudulent utility use, and other damaging consumer actions.
- Determine Personal Behaviour Patterns: energy consumption profiles/patterns in the fine-grained metering data directly or indirectly reveal specific times and locations of electricity use in different locations. Also, the types of activities and appliances can be inferred from such data.
- Determine Specific Appliances Used: the appliances used at specific times can be easily inferred by adversaries if they can access the fine-grained consumption data [40].
- Perform Real-time Surveillance: the utilities collect the fine-grained metering data for energy management and value-added services development. If the time interval becomes shorter, the data collection can be considered as the real-time surveillance by potential adversaries.
- Reveal Activities through Residual Data: the power status of different appliances can reveal such information.
- Target Home Invasions: the living habits of the household can be indicated from the fine-grained metering data. The attackers can easily target a house and learn when the house owners do not stay at home, and then possibly breaks into the house.
- Provide Accidental Invasions: similar to home invasions, criminals may break into houses without target, but learn the living habits of various households.
- Activity Censorship: residential activities could be revealed by the fine-grained metering data. Such information might be shared with local government, law enforcement, or public media. Then, the residents may be under risk of harassment, embarrassment, etc.
- Decisions and Actions based upon Inaccurate Data: PII might be inappropriately modified since metering data are stored, collected, and analysed at different locations.
- Reveal Activities When Used with Data from Other Utilities.

1.7.3 Privacy Laws w.r.t. Smart Grid

In many jurisdictions, privacy laws, which deal with the regulation of personal information of individuals, are considered in the context of individuals' privacy rights and reasonable expectation of privacy. For instance, the United States established Health Insurance Portability and Accountability Act (HIPAA), Financial Service Modernization Act (GLB), Family Educational Rights and Privacy Act (FERPA), etc. The offenders might be prosecuted in a case where individuals' privacy has been compromised. After the Netflix privacy-leaking incident, four customers filed a class action lawsuit against Netflix, alleging that Netflix had violated US fair trade laws and the Video Privacy Protection Act by releasing the datasets (for research and competition purpose). In this section, we introduce some current federal privacy laws w.r.t. smart grid.

1.7.3.1 Smart Meters and the Fourth Amendment [64, 65]

In reality, law enforcements may need to investigate crimes in the houses. They can track residents' daily behaviour and routines using the smart meter data; then, there is no restriction on such data access for law enforcement. By establishing protection of personal privacy rights in investigations, the Fourth Amendment was enacted to restrict access to smart meter data or creating rules to obtain such information. It guarantees that the "right of the people to be secure in their persons, houses, papers and effects, against unreasonable searches and seizures, shall not be violated" [64]. Under the modern conception of the Fourth Amendment, law enforcement officers may not be able to break into system for obtaining the smart metering data when a person has a reasonable expectation of privacy. However, since smart meters are an emerging technology not yet judicially tested, it is difficult to claim the certainty for handling it under the Fourth Amendment [64].

1.7.3.2 Electronic Communications Privacy Act (ECPA) [64]

The ECPA was enacted in 1986 to address the interception of wire, oral, and electronic communications [64]. ECPA prohibits the interception of electronic communications in general, but allows government to conduct surveillance with a specific mechanism (if a party has consented to such interception). In smart grid, the transmission of customers' fine-grained energy consumption via smart grid network falls into the electronic communications under ECPA. Utility would communicate with all the customers and continuously receive information from them via the network (assuming consents from customers have already been established). If the utility consents to interception of the electronic communication by the law enforcement, the surveillance would not violate ECPA. Note that in some types of criminal cases, court orders could authorize electronic surveillance in smart grid without the consent.

1.7.3.3 The Stored Communications Act (SCA)

The SCA (Title II of the ECPA) was enacted in 1986 to address access to stored wire and electronic communications and transactional records [64]. It prohibits unauthorized persons from accessing a facility which provides electronic communication service (ECS). It also limits the ECS providers to disclose information carried or maintained by them. Law enforcement could compel the disclosure of stored communications with a specific mechanism provided by SCA. The protection and disclosure restrictions apply to smart grid (i.e. metering data) since smart meter network might be deployed with the establishment of an ECS.

1.7.3.4 The Federal Privacy Act of 1974 (FPA) [64]

Energy consumption under smart meter is subject to the protections contained in the Federal Privacy Act (FPA). In other words, the FPA protects the smart meter data, and indicates that such time series information is personally identifiable: as a grouping of information of an individual, the smart meter data are typically stored and linked to a consumer's account (may include name, social security number, credit card information, or other PII) [64].

1.7.4 Embedding Privacy Protection into the Design and Implementation of "Smart Grid"

Generating "intelligence" in power grid system, for example, implementing efficient energy distribution, flexible load management, and dynamic pricing model, requires the collection and analysis of huge amount of data in smart grid. Thus, PII might be leaked to untrusted or semi-trusted parties in smart grid. So far, the primary privacy-leaking threats are caused by the fine-grained readings of smart meters in the infrastructure, which are required to monitor the grid status for utilities, consumers, and some other entities. After realizing privacy issues in smart grid infrastructure, contemporary smart grid services start integrating privacy-preserving schemes into their design and implementation [66]. In this section, we outline the privacy-preserving solutions proposed for the design and implementation of smart grid.

1.7.4.1 Metering Data Protection

Smart grid customers concern that their personal information (e.g. their living habits) might be exposed to other parties from the frequently collected metering data. The research question regarding metering data protection is that how to technically anonymise the fine-grained meter readings yet without negatively affecting the network operations, billing applications, and other services. Increasing time intervals of meter readings could clearly remove the attribution of the metering data to specific consumptions; however, many smart grid services might be unavailable for such limited data disclosure. Instead, the following techniques have been proven to be effective for smart metering data protection [66]:

- Anonymization of Metering Data: Separating the technical data (e.g., meter readings) from customer IDs. Thus, the overall meter readings or even the detailed energy consumption cannot be linked to individuals. For this purpose, a third-party ID escrow company should be involved [67].
  Specifically, the utility collects smart meter readings linked to unique IDs instead of customers. In [67], readings are distinguished into two types: (1) low-frequency readings for billing purposes (one reading per week or month, which do not compromise privacy) and (2) high-frequency readings (below a minute). Note that high-frequency readings are required for the maintenance of infra-structure and system, and do not necessarily be linked to the real-world con-sumers. Low-frequency readings can be sent to the utility and billing company, and high-frequency readings should be processed at the next substation (e.g. for load management), but not stored at the utility end. Such work presented a framework that separates two kinds of readings, such that basic billing services are not affected and anonymized metering information can still be used for technical maintenance without compromising the privacy [66].
- Metering Data Obfuscation: Masking the own energy consumption profile with local buffers such as batteries. For instance, with an electric vehicle, the energy consumption of the individual appliances at different times cannot be inferred from the obfuscated data, while the overall consumption remains intact.
- The basic idea of obfuscating the metering data is to locally install intelligent power routers with rechargeable batteries. Then, the usage of individual appliances could be obfuscated. The household load peaks could be smoothened and obscure [68]. The intelligent power management algorithms are used to obfuscate the actual electricity consumption of a household. Varodayan and Khisti [69] presented a preliminary proof that integrating a rechargeable battery and loading/discharging it in non-periodic intervals could greatly reduce the information leakage on the status of the appliances of a household. Note that utilizing a rechargeable battery does not mean that the load peak or energy consumption profile could be completely hidden, but the inference from the metering data could be significantly limited.
  Similarly, Wang et al. [6] proposed a protocol to enable individual meters to report the true energy consumption readings with a predetermined probability. The randomized response model also obfuscated the metering data so as to prevent the inference of individual households' electricity consumption patterns.
- Privacy-Preserving Metering Data Aggregation: Online aggregation of data from geographically colocated consumers. For instance, the utilities can get the aggregated metering information rather than a single household.

- Smart meter data aggregation [70] was originally developed for reducing substantial amount of information and providing aggregated (metering) information for specific purposes. Indeed, metering data aggregation can also reduce the risks of leaking information from the household energy consumption. Two types of aggregation have been realized:

    (1) Spatial aggregation: the metering information is aggregated by geographical locations, where the sum of meter readings of a larger grid segment is transmitted to the data recipients such as the smart grid control centre, instead of the meter readings of single household.
    (2) Temporal aggregation: the aggregation of single readings from a particular meter over a longer interval, which is collected from a single smart meter (e.g. a household). As discussed earlier, the utility of temporally aggregated metering data is limited (e.g. only available for billing purpose).

Aggregation effectively protects privacy but has some new concerns on utility. Skopik raised some possible problems on privacy-preserving metering data aggregation. For instance, for both spatially and temporally aggregated data, it is difficult to run some smart grid services which rely on high-frequency metering information (e.g. dynamic load management, load forecasting, and energy feedback [66]). Also, without the detailed energy consumption information, it is difficult to detect wrong readings or energy theft. Finally, since data should be encrypted before sending out from households for preventing eavesdropping, decryption might be necessary at the other end which performs aggregation operations (e.g. substations). This requires great efforts to implement smart metering/grid services or applications with limited information disclosure.

Note that trade-off between privacy and utility exists in any privacy-preserving technique, including smart grid/metering [40]. Sankar et al. [40] presented a privacy-utility trade-off to quantify privacy and utility requirements of smart meter data. They tried to decouple the revealed meter data from the consumers' personal identifiable information as much as possible with their approach, which distorts the data to minimize the presence of intermittent activity in the data. The trade-off between privacy and utility is quantified based on the rate distortion theory. With an interference-aware reverse waterfilling solution, the privacy–utility tradeoffs on the total load can be achieved, considering the presence of high-power but less private appliance spectra as implicit noise, and filtering out lower-power appliances with a distortion threshold.

1.7.4.2 Privacy-Preserving Applications

Besides the above technical solutions with limited disclosure, cryptographic primitives have been widely utilized to build effective privacy-preserving protocols for many applications in smart grid [46, 71], where efficiency could be relatively ensured. In the following, we introduce some typical examples for this category of privacy-preserving applications in smart grid.

Lin and Fang [72] observed that the aggregated statistics of energy usage could bring intelligence to smart metering-assisted sustainable energy system (e.g. home electricity, water, gas, smart vehicles) and proposed two privacy-preserving schemes to securely collect aggregated statistics while preserving consumers' privacy. The proposed two privacy-preserving schemes are dynamic profiling applications based on the aggregated statistical information of the metering readings: (1) the scheme can extract aggregated statistical information. For example, the scheme enables an aggregator to extract the summation information from the submitted individual responses and can privately answer the statistical question like "What is the total energy consumption when the home temperature is 25 °C?" [72], and (2) extracting correlation information among various factors for the smart system design. For example, the scheme can efficiently answer the query as a conjunction "How many more percent of users consume how much energy on average when annual income is larger than $100K AND the room temperature is 25°C?". Such scheme can also be used as an underlying tool for baseline inference and association rule mining. The system also provides a mechanism to verify the correctness of users' responses which can be deduced from the metering information. The protocols are developed based on the secret key distribution protocol (Diffie–Hellman key-exchange-based protocol).

With the rapid development of smart grid services, vehicle to grid (V2G) becomes an essential component integrated in smart grid network, where the charging status of a battery vehicle should be periodically collected or continuously monitored to perform efficient power scheduling [73, 74]. A battery vehicle is normally associated with a default interest group which is a power grid operator or an organization. In the V2G networks, privacy concerns may arise while providing service in the smart grid system. Yang et al. [74] studied the potential privacy leakage of battery vehicle owners' identity and location and presented a privacy-preserving communication and precise reward architecture, which protects privacy in the process of battery vehicles' monitoring and rewarding. A secure communication architecture based on cryptographic primitives was given to accommodate mutual authentication, confidentiality, data integrity, and privacy protection/anonymity.

Also in the context of V2G network in smart grid, Liu et al. [73] studied the privacy-preserving authentication problem for V2G networks in the smart grid in which every aggregator charges battery vehicles with two modes: home mode and visiting mode. Specifically, battery vehicle may move around in different areas belonging to different groups and thus have requirements on security, privacy, and authentication. The proposed scheme effectively protects the individual privacy while periodically collecting power status data, which refers to a battery vehicle's energy-related status information (e.g. charging efficiency, and battery saturation status). The authors provided a sound security proof for the proposed scheme, including data confidentiality, integrity, availability, mutual authentication, forward/backward security, and privacy preservation.

Summary

In summary, privacy protection is increasingly integrated into the design and implementation of smart grid services, for preventing privacy breach at the individual smart grid component level (end-user, electricity distribution, electricity generation). For the above three components, Wolf [62] illustrated the technologies and applications with privacy issues, e.g. smart meters (remote connect/disconnect of meter, meter detects meter bypass, data collection, communication and storage, in-home appliances that communicate with the utility operator, in-home devices that communicate usage information to the customer, consumer access to energy-related information, and automated feeder equipment), fault detection, load management, and plug-in hybrid electric vehicles. The privacy issues in many of the above applications and technologies have been resolved. However, the privacy-preserving schemes are still worth exploring for the remaining problems by tackling the privacy challenges [2] in the future.

References

1. Li H, Gong S, Lai L, Han Z (2012) Efficient and secure wireless communications for advanced metering infrastructure in smart grids. IEEE Trans Smart Grid 3(3):1540–1551
2. McDaniel P, McLaughlin S (2009) Security and privacy challenges in the smart grid. IEEE Secur Priv 7(3):75–77
3. Bisoi S, Dash AK (2011) The role of utilities in securing a smart grid: electric light and power. Available via http://www.elp.com/articles/print/volume-89/issue-6/sections/the-role-of-utilities-in-securing-a-smart-grid.html. Accessed 6 Jul 2014
4. Wilshusen G (2012) CyberSecurity—challenges in securing the electricity grid. GAO-12926T —Testimony before the Committee on Energy and Natural Resources, US Senate, 17 July 2012
5. ENISA, Smart grid security—annex II. Security aspects of the smart grid. 2012-04-25. https://www.enisa.europa.eu/activities/Resilience-and-CIIP/critical-infrastructure-and-services/smart-grids-and-smart-metering/ENISA_Annex%20II%20-%20Security%20Aspects%20of%20Smart%20Grid.pdf
6. Wang S, Cui L, Que J, Choi D, Jiang X, Cheng S, Xie L (2012) A randomized response model for privacy preserving smart metering. IEEE Trans Smart Grid 3(3):1317–1324
7. Pearson I (2011) Smart grid cyber security for Europe. Energy Policy 39:5211–5218
8. Naone E (2009) Meters for the smart grid: MIT Technology review. September/October 2009:110–111
9. NRG Expert (2011) Chapter 13—Security. Global smart grid report, pp 172–179
10. Steven J, Peterson G, Frinckle D (2010) Smart-grid security issues. IEEE Secur Priv 8(1):81–85
11. Shapiro J (2011) Cyber security and smart grid. In: Presentation at the clean air through energy efficiency (CAFEE) conference, Dallas, 8–11 Nov 2011
12. Aloul F, Al-Ali AR, Al-Dalky R, Al-Mardini M, El-Hajj W (2012) Smart grid security: threats, vulnerabilities, and solutions. Int J Smart Grid Clean Energy 1(1):1–6
13. Echelon (2012) Protect your grid: Echelon's answer for a safe, secure grid. White paper
14. Monticelli A (1999) State estimation in electric power systems: a generalized approach. Springer, Berlin

15. AlMajali A, Viswanathan A, Neuman C (2012) Analyzing resiliency of the smart grid communication architectures under cyber attack. In: Proceedings of the 5th workshop on cyber security experimentation and test, Bellevue, 6 Aug 2012
16. Kosut O, Jia L, Thomas RJ, Tong L (2011) Malicious data attacks on the smart grid. IEEE Trans Smart Grid 2(4):645–658
17. Zhang Z, Gong S, Dimitrovski A, Li H (2013) Time synchronization attack in smart grid: impact and analysis. IEEE Trans Smart Grid 4(1):87–98
18. Lu Z, Lu X, Wang W, Wang C (2010) Review and evaluation of security threats on the communication networks in the smart grid. In: Proceedings of military communications conference, San Jose, 31 Oct–3 Nov 2010
19. Lafferty S, Ghazi T (2011) The increasing importance of security for the smart grid. POWERGrid Int 16(4):60–63
20. Ernst & Young (2011) Attacking the smart grid. Insights on governance, risk and compliance, Dec 2011
21. Ai Ling AP, Masao M (2011) Smart grid information security (IS) functional requirement. Int J Emerg Sci 1(3):371–386
22. Mo Y, Hyun-Jin T, Brancik KK, Dickinson D, Lee H, Perric A, Sinopoli B (2011) Cyber-physical security of a smart grid infrastructure. Proc IEEE 100(1):195–209
23. Zhang Y, Wang L, Sun W, Green RC, Alam M (2011) Distributed intrusion detection system in a multi-layer network architecture of smart grids. IEEE Trans Smart Grid 2(4):796–808
24. Choi K, Chen X, Li S, Kim M, Chae K, Na J (2012) Intrusion detection of MSM based DoS attacks using data mining in smart grid. Energies 5:4091–4109
25. Chen P, Cheng S, Chen K (2012) Smart attacks in smart grid communication networks. IEEE Commun Mag 50(80):24–29
26. Hahn A, Govindarasu M (2011) Cyber attack exposure evaluation framework for the smart grid. IEEE Trans Smart Grid 2(4):835–843
27. Chen T, Sanchez-Aarnoutse JC, Buford J (2011) Petri net modeling of cyber-physical attacks on smart grid. IEEE Trans Smart Grid 2(4):741–749
28. Zonouz S, Rogers K, Berthier R, Bobba R, Sanders W, Overbye T (2012) SCPSE: security-oriented cyber-physical state estimation for power grid critical infrastructures. IEEE Trans Smart Grid 3(4):1790–1799
29. Bobba R, Rogers K, Wang Q, Khurana H, Nahrstedt K, Oberbye T (2010) Detecting false data injection attacks on DC state estimation. In: Proceedings of 1st workshop on secure control systems, Stockholm, Apr 2010
30. Ghansah I (2012) Smart grid cyber security potential threats, vulnerabilities and risks. Public interest energy research (PIER) program interim report, May 2012
31. Li H, Lai L, Zhang W (2011) Communication requirement for reliable and secure state estimation and control in smart grid. IEEE Trans Smart Grid 2(3):476–486
32. Liu Y, Ning P, Reiter M (2011) False data injection attacks against state estimation in electric power grids. ACM Trans Inf Syst Secur 14:13:1–13:33
33. Kim J, Tong L (2013) On topology attacks of a smart grid. IEEE J Sel Areas Commun 31(7):1294–1305
34. Zonouz S, Khurana H, Sanders W, Yardley T (2009) RRE: a game-theoretic intrusion response and recovery engine. IEEE Trans Parallel Distrib Syst 25(2):395–406
35. Locasto M, Wang K, Keromytis A, Stolfo S (2005) FLIPS: Hybrid adaptive intrusion prevention. In: Proceedings symposium on recent advances in intrusion detection, Seattle, pp 82–101, 7–9 Sept 2005
36. Hug G, Giampapa JA (2012) Vulnerability assessment of AC state estimation with respect to false data injection cyber-attacks. IEEE Trans Smart Grid 3(3):1362, 1370
37. Shaw WT (2004) SCADA system vulnerabilities to cyber attack. Electric energy Online 8(6). Retrieved from http://www.electricenergyonline.com/show_article.php?mag=&article=181
38. Kim T (2011) Securing Communication of SCADA Components in Smart Grid Environment. Int J Syst Appl, Eng Dev 5 (2):135–142

39. Skopik F, Ma Z, Bleier T, Gruneis H (2012) A survey on threats and vulnerabilities in smart metering infrastructures. Int J Smart Grid Clean Energy 1(1):22–28
40. Sankar L, Rajagopalan SR, Mohajer S, Poor HV (2012) Smart meter privacy: a theoretical framework. IEEE Trans Smart Grid. doi:10.1109/TSG.2012.2211046
41. Rahman MA, Al-Shaer E, Bera P (2012) A noninvasive threat analyzer for advanced metering infrastructure in smart grid. IEEE Trans Smart Grid. doi:10.1109/TSG.2012.2228283
42. Xiao Z, Xiao Y, Du D (2012) Exploring malicious meter inspection in neighborhood area smart grids. IEEE Trans Smart Grid. doi:10.1109/TSG.2012.2229397
43. Gering K (2010) A meter perspective on cyber security: electronic perspectives. May/June 2010:102–105
44. Bell R (2010) In smart grid security, the details matter: power Grid Internation. Available via: http://www.elp.com/articles/powergrid_international/print/volume-15/issue-4/Features/in-smart-grid-security-the-details-matter.html. Accessed 6 Jul 2014
45. Falk R, Fries S (2011) Smart grid cyber security—an overview of selected scenarios and their security implications. PIK-Praxis der Informationsverarbeitung und Kommunikation 34 (4):168–175
46. Iyer S (2011) Cyber security for smart grid, cryptography, and privacy. Int J Digital Multimedia Broadcast 2011
47. Liu J, Xiao Y, Li S, Liang W, Chen C, Philip L Cyber security and privacy issues in smart grids. IEEE Commun Surv Tutor 14(4):981, 997 (Fourth Quarter)
48. Boyer WF, McBride SA (2009) Study of security attributes of smart grid systems–current cyber security issues. Idaho National Laboratory, USDOE, Under Contract DE-AC07-05ID14517
49. Baumeister T (2010) Literature review on smart grid cyber security. University of Hawaii at Manoa, technical report, 2010
50. Giani A, Bitar E, Garcia M, McQueen M, Khargonekar P, Poolla K (2013) Smart grid data integrity attacks. IEEE Trans Smart Grid 4(3):1244, 1253
51. Gorman S (2009) Electricity grid in U.S. penetrated by spies. Wall St J 8:A1
52. Baldor LC (2010) New threat: hackers look to take over power plants. Associated Press, New York
53. Abur A, Exposito AG (2004) Power system state estimation: theory and implementation. CRC Press, Boca Raton
54. Huang Y, Esmalifalak M, Nguyen H, Zheng R, Han Z, Li H, Song L (2013) Bad data injection in smart grid: attack and defense mechanisms. IEEE Commun Mag 51(1):27–33
55. Liu Y, Reiter MK, Ning P (2009) False data injection attacks against state estimation in electric power grids. In: ACM conference on computer and communications security, pp 21–32
56. Jeu-Min L, Heng-Yau P (2007) A static state estimation approach including bad data detection and identification in power systems. In: IEEE power engineering society general meeting, p 17, June 2007
57. Milli L, Cutsem TV, Pavella MR (1985) Bad data identification methods in power system state estimation, a comparative study. IEEE Trans Power Appar Syst 103(11):3037–3049
58. Rahman MA, Mohsenian-Rad H (2012) False data injection attacks with incomplete information against smart power grids. In: Global communications conference (GLOBECOM), 2012 IEEE, pp 3153–3158
59. Kosut O, Jia L, Thomas RJ, Tong L (2010) Malicious data attacks on smart grid state estimation: attack strategies and countermeasures. In: 1st IEEE international conference on smart grid communications (SmartGridComm), 2010, pp 220, 225, 4–6 Oct 2010
60. Kim TT, Poor HV (2011) Strategic protection against data injection attacks on power grids. IEEE Trans Smart Grid 2(2):326, 333
61. Mohsenian-Rad A-H, Leon-Garcia A (2011) Distributed internet-based load altering attacks against smart power grids. IEEE Trans Smart Grid 2(4):667, 674
62. Cavoukian A, Polonetsky J, Wolf C (2010) Smart privacy for the smart grid: embedding privacy into the design of electricity conservation. Identity Inf Soc 3(2):275–294
63. Rebecca H (2009) SmartGrid privacy concerns. NIST SmartGrid privacy group report, 2009

64. Murrill B, Liu E (2012) Thompson RII Smart meter data: privacy and cybersecurity. CRS Report for Congress, 7-5700, 3 Feb 2012
65. McNeil S (2011) Privacy and the Modern Grid. Harv J Law Technol 25(1)
66. Skopik F (2012) Security is not enough! on privacy challenges in smart grids. Int J Smart Grid Clean Energy 1(1):7–14
67. Efthymiou C, Kalogridis G (2010) Smart grid privacy via anonymization of smart metering data. In: 1st IEEE international conference on smart grid communications (SmartGridComm), pp 238–243. doi:10.1109/SMARTGRID.2010.5622050
68. Kalogridis G, Efthymiou C, Denic SZ, Lewis TA, Cepeda R (2010) Privacy for smart meters: towards undetectable appliance load signatures. In Proceedings of SmartGridComm 2010, pp 232–237
69. David P (2011) Varodayan and Ashish Khisti, "Smart meter privacy using a rechargeable battery: minimizing the rate of information leakage", In Proceedings of ICASSP 2011, pp 1932–1935
70. Kursawe K, Danezis G, Kohlweiss M (2011) Privacy-friendly aggregation for the smart-grid. In Proceedings of international conference on privacy enhancing technologies, pp 175–191
71. Go W, Kwak J (2012) Privacy-enhanced secure data transaction system for smart grid. Int J Secur Appl 6(3):37–44
72. Lin H, Fang Y (2013) Privacy-aware profiling and statistical data extraction for smart sustainable energy systems. IEEE Trans Smart Grid 4(1):332, 340
73. Liu H, Ning H, Zhang Y, Yang LT (2012) Aggregated-proofs based privacy-preserving authentication for V2G networks in the smart grid. IEEE Trans Smart Grid 3(4):1722, 1733
74. Yang Z, Yu S, Lou W, Liu C (2011) Privacy-preserving communication and precise reward architecture for V2G networks in smart grid. IEEE Trans Smart Grid 2(4):697, 706

Chapter 2
Legal Protection of Personal Data in Smart Grid and Smart Metering Systems from the European Perspective

Vagelis Papakonstantinou and Dariusz Kloza

Abstract Smart grids are slowly becoming the future of worldwide energy generation and distribution and they promise, among other things, numerous environmental, and energy efficiency benefits to society. At the same time, however, they are capable of severely invading the inviolability of the most privacy-sensitive place—the home. Therefore, these concerns must be duly taken into consideration while deploying smart grids. This chapter provides an overview, from the European legal perspective, smart grids challenges to the fundamental rights to privacy, personal data protection, and the way Europe has addressed them. It pays special attention to the relevant regulatory requirements and to the means available to properly address these challenges, especially the data protection impact assessment (DPIA). It concludes by a few observations on the efficiency of the European approach.

2.1 Introduction

Smart meters are digital versions of traditional mechanical utility meters that include a two-way communication capacity. They are currently most commonly used for electricity metering, but the principles can be applied to other utilities. These meters can transmit information directly from the metered property to the utility company, potentially in near-real time and with a much higher granularity of data. (By contrast, a traditional meter records the amount of electricity or gas used over a time period and can potentially distinguish between peak and off-peak hours based on a clock). Often, various smart meters in a neighbourhood form a mesh wireless network with a single collection point, which connects to the operating company over a phone line or the Internet. Smart meters are a component of the smart grid, a modernization of electrical infrastructure, with the intended effects of being more responsive to and better able to manage energy demands, and better able to integrate multiple sources of energy. Smart meters are typically the property of the distribution company, not the recipient householder or business. Distribution companies may be different to the electricity retailer, who bills the recipient.

© The Author(s) 2015
S. Goel et al., *Smart Grid Security*, SpringerBriefs in Cybersecurity,
DOI 10.1007/978-1-4471-6663-4_2

The potential benefits for consumers from smart meters include detailed feedback on energy use, potential tips for saving energy, and identification of high-usage or even faulty equipment. The first benefit can be realized by the householders themselves through their own energy meter. Users will be able to understand their household or business uses of energy, compare this with others, programme devices to operate at times of low energy demand, control their expenditure on energy, and take advantage of energy saving plans from their suppliers. Smart devices linked to the smart grid could allow customers to make decisions about heating or other energy use, based upon real-time prices. Smart appliances could be programmed to operate when energy is cheaper (e.g. a dishwasher may run during the middle of the night) or alter their manner of operation (e.g. a thermostat may decrease the heating by a few degrees when there is peak demand for electricity). Smart metering should also facilitate sources of energy that feed back into the grid (e.g. domestic solar panels).

The benefits for the electricity retailers and distributors are significant and include more accurate billing (including tiered time of use pricing), managing credit risks, detecting and managing energy theft, and the potential to better manage electricity demand loads across the network. There are also labour cost savings associated with the end of manual meter reading. Energy supply companies will be able to use the data produced for various research purposes, including testing the efficacy of various demand-response initiatives [1]. Depending on the particular market, the price of wholesale electricity can vary by the hour, half-hour, or quarter hour. Retailers would therefore seek to expose customers to more of this variability in order to encourage demand-reducing behaviour (for example, more selectivity about when to run particular appliances) [2]. The ability to remotely shift customers to prepayment plans in case of default and the ease of changing account holders offers operational cost savings to utility companies.

In view of their potential substantial benefits, the roll-out of smart grids has been raised as a priority for the European Union (EU), which aims at having 80 % of consumers with smart metering systems in place by 2020. To this end, in parallel to other initiatives,[1] the EU has released a series of regulatory texts of varying statuses that are aimed at encouraging their implementation across the EU while also setting the basic end-user protection rules with regard to their use. Certain EU Member States, for instance the United Kingdom [3] or the Netherlands, have also been active in the field, elaborating both on smart grid implementation and their ethical, privacy, data protection, and security ramifications for the individuals.[2] The same is also apparently true in the United States [4].

The EU law shall form the legal framework within which smart grid security issues shall be assessed. Indeed, in the event of a security breach, there exist two types of conceivable infringements: one leading to the loss or unauthorized access and use of personal data and the other leading to some type of fraud. The analysis that follows, however, shall only focus on the former; fraud as a result of smart grid

---

[1] See, for instance, the European Commission's Smart Grids Task Force. Cf. *infra*, at Sect. 2.4.2.
[2] Cf. *infra*, at 2.10.

security breaches will have to be assessed, first, once actual smart grid implementations are in place and, second, under the penal law provisions of each Member State (that may differ substantially).

Therefore, data protection legislation constitutes the legal framework that is apparently directly affected by smart grid and smart metering systems implementations. After all, it is for this reason that the relevant analysis, as it will be seen, has already attracted significant attention that is perhaps even disproportionate to the level of use such systems have found across the EU.

This part of the book aims to analyse the challenges that smart grids and smart metering systems pose to the protection of privacy and personal data. It takes predominantly a legal perspective. To that end, the authors have chosen the European viewpoint as an "exemplar". Despite the focus on data protection, however, in order to give a complete picture, the deliberations on personal data protection are preceded by some background information concerning smart grids and smart metering systems.

The authors first analyse the EU action concerning smart grids and smart metering system, i.e. the regulatory framework thereof, policy initiatives as well as relevant stakeholder in the field (Sects. 2.2–2.4). Second, we embark on the analysis of the general data protection framework in the European Union (Sect. 2.5). Third, we continue with the analysis of the interaction of smart grids and smart metering systems with the protection of personal data (Sects. 2.6–2.7). This is followed by an overview of the so-called privacy and personal data protection "tools" that might prove useful for the operators of smart grids and smart metering systems (Sect. 2.8). Next, in Sect. 2.9, we discuss the interests of the consumer. In Sect. 2.10, we briefly mention two examples of the national implementations of the smart grids and smart metering systems, i.e. the Netherlands and the United Kingdom. We conclude, in Sect. 2.11, with observations concerning protecting personal data in smart grids and smart metering systems.

This part of the book is structured predominantly as texts and materials. This book has been written on the basis of the law as it stood on 1 November 2014.

2.2 The Rationale and *Modus Operandi* for the EU Action Concerning Smart Grid and Smart Metering Systems

For the sake of clarity, it is important, first, to explain the rationale and *modus operandi* of the involvement of the EU in the deployment and regulation of smart grid and smart metering systems.

The key to understand this phenomenon is at least threefold. The first reason has to do with the rationale of European integration. From the historical viewpoint, the European integration was launched in early 1950s with a functional interest in *energy*, i.e. in the supranational governance of both the production and usage of coal and steel[3]

---

[3] Cf. Treaty establishing the European Coal and Steel Community (Paris 1951).

and, subsequently, of nuclear energy.[4] The integration of these two industry sectors had been considered a means to achieve the larger goals of peace and prosperity, which was particularly important in the post-war Western Europe. Further, mid-1950s generated a political agreement to move in the direction of a broader *economic* integration [5] and resulted in the establishment of the European Economic Community (EEC)[5] in 1958, a forerunner of the contemporary EU.[6] Nowadays, some 50 years later, the European integration to a large extent is still driven by economic reasons and one of the primary aims thereof is the development of the internal market. The EU Treaties[7] define the internal market as an area "without internal frontiers in which the free movement of goods, persons, services and capital is ensured".[8]

The second reason has to do with the integration of national energy markets. It was not until 1980s when the EU has become increasingly interested in developing an integrated energy market, having realized that "the energy sector should not be isolated from the internal market but should be subject to the liberalisation policies that affect other sectors" [6, 343]. The 1980s also saw Member States preference switched from essentially national solutions to the quest for supranational ones. Nugent [6, 343] further argues this development has been stimulated by factors such as:

- the centrality of energy to any modern economy,
- immense savings accruing from an integrated energy market,
- growing recognition of the over-reliance of the EU on external suppliers,
- the "aggressive" stances of some EU energy suppliers,
- the need to tackle climate change, to save energy, and to promote clearer energy production.

The third reason has to do with the intertwining energy policy with other policies. This progressive development made the EU energy polices intertwined with other relevant policies, such a climate change and environment. With regard to the last one, Art 11 TFEU explicitly states that "environmental protection requirements must be integrated into the definition and implementation of the Union's policies and activities". The said provision, introduced by the Lisbon Treaty (2007), is a codification of a practice known from late 1990s as the Cardiff process of integrating environmental considerations into the work of all policy sectors [7, 367].[9]

In consequence, the EU energy policy was born and matured with a focus on [6, 343]:

---

[4] Cf. Treaty establishing the European Atomic Energy Community (Rome 1957).

[5] Cf. Treaty establishing the European Economic Community (Rome 1957).

[6] Cf. Treaty on European Union (Maastricht 1992).

[7] Currently, the EU is based on two basic international agreements defining the constitutional order of the Union: the Treaty on European Union (TEU) and the Treaty on the Functioning of the European Union (TFEU). These Treaties undergo a numerous amendments since their first inception as the Treaties of Rome (1957) and the Treaty of Maastricht (1992). The Treaty of Lisbon (2007) constitutes the most recent amendment to the EU Treaties.

[8] Art 26(1) TFEU.

[9] Cf. http://ec.europa.eu/environment/integration/integration.htm.

- developing an internal market in energy,
- developing external energy relations and ensuring security of supply,
- managing demand,
- diversifying sources,
- minimizing the negative impact on the environment of energy use and production,
- combating the climate change.

To a large extent, these goals have been codified in the EU Treaties:

Treaty on the Functioning of the European Union (1957, revised 2009)

Art 194
1. In the context of the establishment and functioning of the internal market and with regard for the need to preserve and improve the environment, Union policy on energy shall aim, in a spirit of solidarity between Member States, to:

  (a) ensure the functioning of the energy market;
  (b) ensure security of energy supply in the Union;
  (c) promote energy efficiency and energy saving and the development of new and renewable forms of energy; and
  (d) promote the interconnection of energy networks.

From the formal point of view, when the common energy policy gradually got prominence among all the EU policies, it has become the so-called *shared competence*. This means that both the EU and its Member States may regulate in given areas, yet the Member States can exercise their competence to the extent that the Union has not done so or the EU ceased to exercise it.[10]

In case of shared competences, the extent of the involvement of the EU is governed by the principle of *subsidiarity*. This means that "the Union shall act only if and in so far as the objectives of the proposed action cannot be sufficiently achieved by the Member States, either at central level or at regional and local level, but can rather, by reason of the scale or effects of the proposed action, be better achieved at Union level".[11] The extent of the EU action is further limited by the principle of *proportionality*, which means that "the content and form of Union action shall not exceed what is necessary to achieve the objectives".[12]

---

[10] Art 2(4) TFEU.
[11] Art 5(3) TEU.
[12] Art 5(4) TEU.

2.3 The EU Regulatory Framework for Smart Grid and Smart Metering Systems

2.3.1 The Legally Binding Framework

Secondly, it is important to overview the *general* regulatory framework for smart grid and smart metering systems in the EU.

Given the objectives of the EU energy policy (cf. *supra*, at 2.2), supplemented by the goals of developing the internal market and protecting the environment, among others, the EU enacted a number of legally binding instruments—predominantly directives[13]—that regulate the deployment of smart grid and smart metering systems. They focus largely on the conditions for their deployment (e.g. 80 % deployment by 2020)[14] and on the functional requirements thereof (e.g. information on actual—as opposed to estimated—energy consumption).[15]

2.3.1.1 Measuring Instruments Directive (2004)

From the historical perspective, the first legally binding instrument mentioning smart grid and smart metering systems was the so-called Measuring Instruments Directive (2004).[16] The directive applies to measuring instruments for water, gas, electricity or heat. First, it establishes the essential requirements that these instruments will have to satisfy and the conformity assessment that they have to undergo prior to their deployment and putting into use. Second, it provides that Member States shall not impede the placing on the market and putting into use of any measuring instrument that carries the CE conformity marking and supplementary metrology marking.

Important for our purposes is a fact that this Directive implicitly prescribes the minimum period of the information retention within an electricity meter:[17]

---

[13] For the sake of clarity, the EU has a power to enact binding legislative instruments of two main types. A directive binds the Member States as to the goals but leaves the means of implementation to them. Thus, a directive is always implemented into a national legal system, usually by an act of parliament. A regulation is a directly binding instrument and requires no implementation in a national legal system. These two types of legal instruments are supplemented by non-binding ones such as recommendations and opinions. Various instruments will often be used in conjunction with each other. For more information on the EU legislative toolbox, cf. [5, 111–117].

[14] Cf. *infra*, at 2.3.1.2.

[15] Cf. *infra*, at 2.3.1.3.

[16] Directive 2004/22/EC of the European Parliament and of the Council of 31 March 2004 on measuring instruments, OJ L 135, 30.4.2004, pp. 1–80. All EU legislation can be accessed via http://eur-lex.europa.eu.

[17] Annex MI-003, paragraph 5(3).

In the event of loss of electricity in the circuit, the amounts of electrical energy measured shall remain available for reading during a period of at least 4 months.

2.3.1.2   Third Energy Package (2009)

With a view to "make the energy market fully effective" and create a genuine "single EU gas and electricity market",[18] the 2009 Third Energy Package brought further integration of internal energy market. The Package consists of five main legal instruments:

- The Electricity Internal Market Directive,[19]
- The Gas Internal Market Directive,[20]
- The Network for Cross-border Exchanges in Electricity Regulation,[21]
- The Natural Gas Transmission Networks Regulation,[22]
- The ACER (Agency for the Cooperation of Energy Regulators) Regulation.[23]

The Electricity Internal Market Directive encourages the "modernisation of distribution networks, such as through the introduction of smart grids, which should be built in a way that encourages decentralised generation and energy efficiency".[24] In order to "promote energy efficiency, Member States ... shall strongly recommend that electricity undertakings optimise the use of electricity, for example by ... introducing intelligent metering systems or smart grids, where appropriate".[25]

---

[18] European Commission, *Questions and Answers on the third legislative package for an internal EU gas and electricity market*, MEMO 11/125, Brussels, 2 March 2011. http://europa.eu/rapid/press-release_MEMO-11-125_en.htm.

[19] Directive 2009/72/EC of the European Parliament and of the Council of 13 July 2009 concerning common rules for the internal market in electricity and repealing Directive 2003/54/EC, OJ L 211, 14.8.2009, pp. 55–93.

[20] Directive 2009/73/EC of the European Parliament and of the Council of 13 July 2009 concerning common rules for the internal market in natural gas and repealing Directive 2003/55/EC, OJ L 211, 14.8.2009, pp. 94–136.

[21] Regulation (EC) No 714/2009 of the European Parliament and of the Council of 13 July 2009 on conditions for access to the network for cross-border exchanges in electricity and repealing Regulation (EC) No 1228/2003, OJ L 211, 14.8.2009, pp. 15–35.

[22] Regulation (EC) No 715/2009 of the European Parliament and of the Council of 13 July 2009 on conditions for access to the natural gas transmission networks and repealing Regulation (EC) No 1775/2005, OJ L 211, 14.8.2009, pp. 36–54.

[23] Regulation (EC) No 713/2009 of the European Parliament and of the Council of 13 July 2009 establishing an Agency for the Cooperation of Energy Regulators, OJ L 211, 14.8.2009, pp. 1–14.

[24] Recital 27.

[25] Art 3(11).

The Directive conditions the roll-out of smart grid and smart metering systems to the positive economic assessment "of all the long-term costs and benefits to the market and the individual consumer".[26] In case "roll-out of smart meters is assessed positively, at least 80 % of consumers shall be equipped with intelligent metering systems by 2020".[27]

The Directive also touches upon the processing of personal data within electricity meters. In particular:

- the regulatory authority shall ensure "access to customer consumption data",[28]
- the consumer shall "have at their disposal their consumption data, and shall be able to, by explicit agreement and free of charge, give any registered supply undertaking access to its metering data",[29]
- the consumer shall be "properly informed of actual electricity consumption and costs frequently enough to enable them to regulate their own electricity consumption. That information shall be given by using a sufficient time frame, which takes account of the capability of customer's metering equipment and the electricity product in question. Due account shall be taken of the cost-efficiency of such measures",[30]
- the consumer shall have a right to a contract with their electricity service provider that "specifies information relating to consumer rights, including on the complaint handling and all of the information referred to in this point, clearly communicated through billing or the electricity undertaking's web site",[31]
- No additional costs shall be charged to the consumer for any of the above-mentioned services.[32]

2.3.1.3 New Energy Efficiency Directive (2012)

One of the focuses of the EU energy policy is the efficiency goals. As defined by the New Energy Efficiency Directive (2012),[33] the main objectives are:

---

[26] Recital 55 and Annex 1, paragraph 2.

[27] Annex 1, paragraph 2.

[28] Art 37(1)(p).

[29] Annex I, paragraph 1(h).

[30] Annex I, paragraph 1(i).

[31] Annex I, paragraph 1(a).

[32] Annex I, paragraphs 1(h)–1(j).

[33] Directive 2012/27/EU of the European Parliament and of the Council of 25 October 2012 on energy efficiency, amending Directives 2009/125/EC and 2010/30/EU and repealing Directives 2004/8/EC and 2006/32/EC, OJ L 315, 14.11.2012, pp. 1–56.

Recital 1

The Union is facing unprecedented challenges resulting from increased dependence on energy imports and scarce energy resources, and the need to limit climate change and to overcome the economic crisis. Energy efficiency is a valuable means to address these challenges. It improves the Union's security of supply by reducing primary energy consumption and decreasing energy imports. It helps to reduce greenhouse gas emissions in a cost-effective way and thereby to mitigate climate change. Shifting to a more energy-efficient economy should also accelerate the spread of innovative technological solutions and improve the competitiveness of industry in the Union, boosting economic growth and creating high quality jobs in several sectors related to energy efficiency.

The new Directive replaced the analogous instrument from 2006. Among other novelties, the new directive sets forth further detailed and specific functional requirements of smart meters as a function of empowering "final customers as regards access to information from the metering and billing of their individual energy consumption".[34]

The Energy Efficiency Directive (2012)

Article 9
Metering

1. Member States shall ensure that, in so far as it is technically possible, financially reasonable and proportionate in relation to the potential energy savings, final customers for electricity, natural gas, district heating, district cooling and domestic hot water are provided with competitively priced individual meters that accurately reflect the final customer's actual energy consumption and that provide information on actual time of use. [...]

2. Where, and to the extent that, Member States implement intelligent metering systems and roll out smart meters for natural gas and/or electricity in accordance with Directives 2009/72/EC and 2009/73/EC:[35]

   (a) they shall ensure that the metering systems provide to final customers information on actual time of use and that the objectives of energy efficiency and benefits for final customers are fully taken into account when establishing the minimum functionalities of the meters and the obligations imposed on market participants;
   (b) they shall ensure the security of the smart meters and data communication, and the privacy of final customers, in compliance with relevant Union data protection and privacy legislation;
   (c) in the case of electricity and at the request of the final customer, they shall require meter operators to ensure that the meter or meters can account for electricity put into the grid from the final customer's premises;

---

[34] Recital 33.

[35] Electricity Internal Market Directive and Gas Internal Market Directive, respectively. cf. *supra*, at 2.3.1.2 [VP & DK].

(d) they shall ensure that if final customers request it, metering data on their electricity input and off-take is made available to them or to a third party acting on behalf of the final customer in an easily understandable format that they can use to compare deals on a like-for-like basis;

(e) they shall require that appropriate advice and information be given to customers at the time of installation of smart meters, in particular about their full potential with regard to meter reading management and the monitoring of energy consumption.

Article 10
Billing information

2. Meters installed in accordance with Directives 2009/72/EC and 2009/73/EC shall enable accurate billing information based on actual consumption. Member States shall ensure that final customers have the possibility of easy access to complementary information on historical consumption allowing detailed self-checks.

Complementary information on historical consumption shall include:

(a) cumulative data for at least the three previous years or the period since the start of the supply contract if this is shorter. The data shall correspond to the intervals for which frequent billing information has been produced; and

(b) detailed data according to the time of use for any day, week, month and year. These data shall be made available to the final customer via the internet or the meter interface for the period of at least the previous 24 months or the period since the start of the supply contract if this is shorter.

3. Independently of whether smart meters have been installed or not, Member States:

(a) shall require that, to the extent that information on the energy billing and historical consumption of final customers is available, it be made available, at the request of the final customer, to an energy service provider designated by the final customer;

(b) shall ensure that final customers are offered the option of electronic billing information and bills and that they receive, on request, a clear and understandable explanation of how their bill was derived, especially where bills are not based on actual consumption;

(c) shall ensure that appropriate information is made available with the bill to provide final customers with a comprehensive account of current energy costs, in accordance with Annex VII;

(d) may lay down that, at the request of the final customer, the information contained in these bills shall not be considered to constitute a request for payment. In such cases, Member States shall ensure that suppliers of energy sources offer flexible arrangements for actual payments;

(e) shall require that information and estimates for energy costs are provided to consumers on demand in a timely manner and in an easily understandable format enabling consumers to compare deals on a like-for-like basis.

Article 11
Cost of access to metering and billing information

1. Member States shall ensure that final customers receive all their bills and billing information for energy consumption free of charge and that final customers also have access to their consumption data in an appropriate way and free of charge. [...]

Article 12
Consumer information and empowering programme

1. Member States shall take appropriate measures to promote and facilitate an efficient use of energy by small energy customers, including domestic customers. These measures may be part of a national strategy.

2. For the purposes of paragraph 1, these measures shall include one or more of the elements listed under point (a) or (b):

(a) a range of instruments and policies to promote behavioural change which may include:

   (i)   fiscal incentives;
   (ii) access to finance, grants or subsidies;
   (iii) information provision;
   (iv) exemplary projects;
   (v)   workplace activities;

(b) ways and means to engage consumers and consumer organisations during the possible roll-out of smart meters through communication of:

   (i)   cost-effective and easy-to-achieve changes in energy use;
   (ii) information on energy efficiency measures.

Article 17
Information and training

1. Member States shall ensure that information on available energy efficiency mechanisms and financial and legal frameworks is transparent and widely disseminated to all relevant market actors, such as consumers, builders, architects, engineers, environmental and energy auditors, and installers of building elements as defined in Directive 2010/31/EU.[36]

Member States shall encourage the provision of information to banks and other financial institutions on possibilities of participating, including through the creation of public/private partnerships, in the financing of energy efficiency improvement measures.

2. Member States shall establish appropriate conditions for market operators to provide adequate and targeted information and advice to energy consumers on energy efficiency.

3. The Commission shall review the impact of its measures to support the development of platforms, involving, inter alia, the European social dialogue bodies in fostering training programmes for energy efficiency, and shall bring forward further measures if appropriate. The Commission shall encourage European social partners in their discussions on energy efficiency.

4. Member States shall, with the participation of stakeholders, including local and regional authorities, promote suitable information, awareness-raising and training initiatives to inform citizens of the benefits and practicalities of taking energy efficiency improvement measures.

5. The Commission shall encourage the exchange and wide dissemination of information on best energy efficiency practices in Member States.

---

[36] Directive 2010/31/EU of the European Parliament and of the Council of 19 May 2010 on the energy performance of buildings, OJ L 153, 18.6.2010, pp. 13–35 [VP & DK].

Article 25
Online platform

The Commission shall establish an online platform in order to foster the practical implementation of this Directive at national, regional and local levels. That platform shall support the exchange of experiences on practices, benchmarking, networking activities, as well as innovative practices. [...]

ANNEX VII

Minimum requirements for billing and billing information based on actual consumption

1. *Minimum requirements for billing*

1.1. *Billing based on actual consumption*

In order to enable final customers to regulate their own energy consumption, billing should take place on the basis of actual consumption at least once a year, and billing information should be made available at least quarterly, on request or where the consumers have opted to receive electronic billing or else twice yearly. Gas used only for cooking purposes may be exempted from this requirement.

1.2. *Minimum information contained in the bill*

Member States shall ensure that, where appropriate, the following information is made available to final customers in clear and understandable terms in or with their bills, contracts, transactions, and receipts at distribution stations:

(a) current actual prices and actual consumption of energy;
(b) comparisons of the final customer's current energy consumption with consumption for the same period in the previous year, preferably in graphic form;
(c) contact information for final customers' organisations, energy agencies or similar bodies, including website addresses, from which information may be obtained on available energy efficiency improvement measures, comparative end-user profiles and objective technical specifications for energy-using equipment.

In addition, wherever possible and useful, Member States shall ensure that comparisons with an average normalised or benchmarked final customer in the same user category are made available to final customers in clear and understandable terms, in, with or signposted to within, their bills, contracts, transactions, and receipts at distribution stations.

1.3. *Advice on energy efficiency accompanying bills and other feedback to final customers*

When sending contracts and contract changes, and in the bills customers receive or through websites addressing individual customers, energy distributors, distribution system operators and retail energy sales companies shall inform their customers in a clear and understandable manner of contact information for independent consumer advice centres, energy agencies or similar institutions, including their internet addresses, where they can obtain advice on available energy efficiency measures, benchmark profiles for their energy consumption and technical specifications of energy using appliances that can serve to reduce the consumption of these appliances.

2.3.2 The Non-Binding Framework

The patchwork of legally binding instruments regulating smart grid and smart metering systems in the EU (cf. *supra*, at 2.3.1) is supplemented by a number of non-binding policy instruments, such as recommendations and opinions, issued by various EU institutions and bodies.

The roll-out of smart grids and smart metering systems in Europe has been embodied into key policy initiatives of the EU. Mentioned only in a recent flagship innovation initiative of the EU, the Digital Agenda for Europe (2010),[37] the governance of smart grids and smart metering systems have been for the first time comprehensively addressed in a European Commission's policy document "Smart Grids: from innovation to deployment" (2010). Subsequently, their deployment received political backing from the European Council (2011). In 2012, the EU issued guidelines for cost-benefit analysis of smart grid projects.

Commission Communication: *A Digital Agenda for Europe* (2010)[38]

2.7.1. ICT for environment

The EU has committed to cutting its greenhouse gas emissions by at least 20 % by 2020 compared to 1990 levels and to improving energy efficiency by 20 %. The ICT sector has a key role to play in this challenge:

- ICT offer potential for a structural shift to less resource-intensive products and services, for energy savings in buildings and electricity networks, as well as for more efficient and less energy consuming intelligent transport systems;
- The ICT sector should lead the way by reporting its own environmental performance by adopting a common measurement framework as a basis for setting targets to reduce energy use and greenhouse gas emissions of all processes involved in production, distribution, use and disposal of ICT products and delivery of ICT services.

Cooperation between the ICT industry, other sectors and public authorities is essential to accelerate development and wide-scale roll out of ICT-based solutions for smart grids and meters, near-zero energy buildings and intelligent transport systems. It is essential to empower individuals and organisations with information that will help them to reduce their own carbon footprint. The ICT sector should deliver modelling, analysis, monitoring and visualisation tools to evaluate the energy performance and emissions of buildings, vehicles, companies, cities and regions. Smart grids are essential for the move to a low carbon economy. They will enable active control of transmission and distribution via advanced ICT infrastructure communication and control platforms. For the different grids to work together efficiently and safely, open transmission-distribution interfaces will be needed. [...]

---

[37] Communication from the Commission to the European Parliament, the Council, the European Economic and Social Committee and the Committee of the Regions, *A Digital Agenda for Europe*, Brussels, 26 August 2010, COM (2010) 245 final/2.

[38] Cf. *supra*, note 37.

ACTIONS

The Commission will [...]

- Assess by 2011 the potential contribution of smart grids to the decarbonisation of energy supply in Europe and define a set of minimum functionalities to promote the interoperability of Smart Grids at European level by the end of 2010. [...]

Member States should:

- Agree by the end of 2011 common additional functionalities for smart meters [...].

Commission Communication: *Smart Grids: from innovation to deployment* **(2010)**[39]

2.2. Addressing data privacy and security issues

Developing legal and regulatory regimes that respect consumer privacy in cooperation with the data protection authorities, in particular with the European Data Protection Supervisor, and facilitating consumer access to and control over their energy data processed by third parties is essential for the broad acceptance of Smart Grids by consumers. Any data exchange must also protect the sensitive business data of grid operators and other players, and enable companies to share Smart Grids data in a secure way.

Directive 95/46/EC on the protection of personal data constitutes the core legislation governing the processing of personal data. The Directive is technology-neutral and the data processing principles apply to the processing of personal data in any sector, so also cover some Smart Grids aspects. The definition of personal data is particularly relevant, as the distinction between personal and non-personal data is of outmost importance for further Smart Grids deployment. If the data processed are technical and do not relate to an iden-tified or identifiable natural person, then Distributed System Operators (DSOs), smart meter operators and energy service companies could process such data without needing to seek prior consent from grid users. While the European data framework is appropriate and does not need to be extended, some adaptations might be needed in the specific national legal frameworks in order to accommodate some Smart Grids foreseen functionalities. With the wide deployment of Smart Grids, the obligation to notify national data protection author-ities of the processing of personal data is naturally likely to increase. Member States will have to ensure, when setting up Smart Grids and more particularly when deciding on the division of roles and responsibilities regarding ownership, possession and access to data, that this is done in full compliance with the EU and national data protection legislation.

The Smart Grids Task Force has agreed that a 'privacy by design' approach is needed. This will be integrated in the standards being developed by the ESOs.

Finally, developing and maintaining a secure network is essential for continuity of resources and the safety of consumers. It is important to ensure the security and resilience of the infrastructures supporting Smart Grids deployment in Europe. [...]

---

[39] Communication from the Commission to the European Parliament, the Council, the European Economic and Social Committee and the Committee of the Regions, *Smart Grids: from innovation to deployment*, Brussels, 12 April 2011, COM(2011) 202 final.

2. Actions on data privacy and security of data in Smart Grids

- The Commission will monitor the provisions of national sectoral legislation that might apply to take into account the data protection specificities of Smart Grids.
- The ESOs will develop technical standards for Smart Grids taking the 'privacy by design' approach.
- The Commission will continue bringing together the energy and ICT communities within an expert group to assess the network and information security and resilience of Smart Grids as well as to support related international cooperation.

Conclusions of the European Council of 4 February 2011[40]

4. The internal market should be completed by 2014 so as to allow gas and electricity to flow freely. This requires in particular that in cooperation with ACER national regulators and transmission systems operators step up their work on market coupling and guidelines and on network codes applicable across European networks. Member States, in liaison with European standardization bodies and industry, are invited to accelerate work with a view to adopting technical standards for electric vehicle charging systems by mid-2011 and for smart grids and meters by the end of 2012. The Commission will regularly report on the functioning of the internal energy market, paying particular attention to consumers including the more vulnerable ones in line with the Council conclusions of 3 December 2010. [...]

10. The EU and its Member States will promote investment in renewables and safe and sustainable low carbon technologies and focus on implementing the technology priorities established in the European Strategic Energy Technology plan. The Commission is invited to table new initiatives on smart grids, including those linked to the development of clean vehicles, energy storage, sustainable bio fuels and energy saving solutions for cities.

The Commission Recommendation on preparation for the roll-out of smart metering[41] constitutes the core instrument in that regard. The recommendation is a comprehensive instrument that addresses three main concerns:

(a) personal data protection and security (§§ 4–29), which will be discussed in detail *infra*, at 2.7;
(b) economic assessment of the long-term costs and benefits (§§ 30–38);
(c) common minimum functional requirements for smart meters (§§ 39–42), cf. Exhibit 1.

---

[40] Cf. https://www.consilium.europa.eu/uedocs/cms_data/docs/pressdata/en/ec/119175.pdf.

[41] Commission Recommendation of 9 March 2012 on preparations for the roll-out of smart metering, COM(2012) 1342 final, 2012/148/EU, OJ L 73, 13.3.2012, pp. 9–22; *hereinafter*: the 2012 Recommendation.

Exhibit 1: Common minimum functional requirements for smart meters

1. For the customer

  (a) Provide readings directly to the customer and any third party designated by the consumer
  (b) Update the readings referred to in point (a) frequently enough to allow the information to be used to achieve energy savings

2. For the metering operator

  (c) Allow remote reading of meters by the operator
  (d) Provide two-way communication between the smart metering system and external networks for maintenance and control of the metering system
  (e) Allow readings to be taken frequently enough for the information to be used for network planning

3. For commercial aspects of energy supply

  (f) Support advanced tariff systems
  (g) Allow remote on/off control of the supply and/or flow or power limitation

4. For security and data protection

  (h) Provide secure data communications
  (i) Fraud prevention and detection

5. For distributed generation

  (j) Provide import/export and reactive metering

Guidelines for conducting a cost-benefit analysis of Smart Grid projects (2012)[42]

5.2. Externalities and social impact

Apart from addressing the deployment merit of a project, the qualitative analysis should granularly identify and assess all costs and benefits that spill over from the project into society and that cannot be monetised and included in the economic analysis (externalities). All externalities should be listed and expressed in physical terms (e.g. use decibels to quantify noise reduction benefit). [...]

Social impacts represent a significant portion of the possible externalities of a Smart Grid project. It is expected that society at large may benefit from the Smart Grid through the resulting improvement in areas like national security, environmental conditions, public health or economic growth [...] Although difficult to monetise, the social impact of Smart Grid implementation is significant. These benefits are complex to evaluate, but understanding their importance is essential for grasping the (entire) value of Smart Grids. Therefore, in the remainder of this section, we will present some of the areas worth considering in the assessment of the social impact of a Smart Grid project. [...]

---

[42] European Commission, Joint Research Centre, Institute for Energy and Transport, *Guidelines for conducting a cost-benefit analysis of Smart Grid projects*, Report EUR 25246 EN, Petten 2012. http://ses.jrc.ec.europa.eu/sites/ses.jrc.ec.europa.eu/files/publications/guidelines_for_conducting_a_cost-benefit_analysis_of_smart_grid_projects.pdf.

Social acceptance

In several instances, social acceptance is key to the successful implementation of Smart Grid projects. Social resistance may arise due to concerns over transparency, fair benefit sharing or environmental impact. If applicable, an assessment of the level of social resistance (or participation) to the project should be presented, including a description of the means adopted to ensure social acceptance and their effectiveness. […]

Privacy and security

This analysis should address the foreseeable activities in developing measures to ensure data privacy and cyber-security. It may qualitatively include the additional costs estimated for implementing preventive measures. […]

IV. PERFORMANCE ASSESSMENT, EXTERNALITIES AND SOCIAL IMPACT

Guideline 10—Qualitative impact analysis: non-monetary appraisal

The CBA should be complemented by a qualitative impact analysis, i.e. a qualitative estimation of additional costs and benefits that cannot be monetised and included in a CBA. The qualitative impact analysis should include (1) deployment merit of the project (performance assessment); (2) externalities, with particular reference to social impacts. […]

Externalities and social impacts

[…] Social impacts typically represent a significant portion of the project externalities. Some areas of focus include:

- job impact
- safety
- environmental impact
- social acceptance
- time lost/saved by consumers
- enabling new services and applications and market entry to third parties
- reduction of the gap in skills and personnel
- privacy and security. […]

ANNEX IV—KEY PERFORMANCE INDICATORS AND BENEFITS

Create a market mechanism for new energy services such as energy efficiency or energy consulting for customers

46. 'Simple' and/or automated changes to consumers' energy consumption in reply to demand/response signals are enabled
47. Data ownership is clearly defined and data processes in place to allow for service providers to be active with customer consent
48. Physical grid-related data are available in an accessible form
49. Transparency of physical connection authorisation, requirements and charges
50. Effective consumer complaint handling and redress. This includes clear lines of responsibility should things go wrong

2.4 Actors in the Field of Energy Regulation in the EU

2.4.1 European Commission—Directorate-General for Energy (DG ENER)

<u>Mission statement of DG Energy</u>[43]

[The] Directorate-General for Energy is responsible for developing and implementing a European energy policy. Through the development and implementation of innovative policies, the Directorate-General aims at:

- Contributing to setting up an energy market providing citizens and business with affordable energy, competitive prices and technologically advanced energy services.
- Promoting sustainable energy production, transport and consumption in line with the EU 2020 targets and with a view to the 2050 decarbonisation objective.
- Enhancing the conditions for secure energy supply in a spirit of solidarity between Member States.

In developing a European energy policy, the Directorate-General aims to support the Europe 2020 programme which, for energy, is captured in the Energy 2020 strategy.[44]

The Directorate-General carries out its tasks in many different ways. For example, it develops strategic analyses and policies for the energy sector; promotes the completion of the internal energy market encompassing electricity, gas, oil and oil products, solid fuels and nuclear energy; supports the reinforcement of energy infrastructure, ensures that indigenous energy sources are exploited in safe and competitive conditions; ensures that markets can deliver agreed objectives, notably in efficiency and renewable energies; promotes and conducts an EU external energy policy; facilitates energy technology innovation; develops the most advanced legal framework for nuclear energy, covering safety, security and non-proliferation safeguards; monitors the implementation of existing EU law and makes new legislative proposals; encourages the exchange of best practices and provides information to stakeholders.

All this work is aided by expert input from the Executive Agency for Competitiveness and Innovation (EACI),[45] the Euratom Supply Agency (ESA)[46] and the Agency for the Cooperation of Energy Regulators (ACER, operational from March 2011).[47]

---

[43] Cf. http://ec.europa.eu/dgs/energy/mission_en.htm.

[44] Cf. http://ec.europa.eu/energy/strategies/2010/2020_en.htm [VP & DK].

[45] Now: Executive Agency for Small and Medium-sized Enterprises (EASME), http://ec.europa.eu/easme [VP & DK].

[46] Cf. http://ec.europa.eu/euratom [VP & DK].

[47] Cf. *infra*, at 2.4.3 [VP & DK].

2.4.2 Smart Grids Task Force (SGTF)

The European Commission set up the Smart Grids Task Force (SGTF) at the end of 2009.[48] The mission thereof is to advise the Commission on policy and regulatory frameworks at European level to coordinate the first steps towards the implementation of smart grids and smart metering systems under the provision of the Third Energy Package.[49]

The Task Force consists of the Steering Committee (SC) and four expert groups (EG):

- EG1: Reference Group for Smart Grid Standards,
- EG2: Expert Group for Regulatory Recommendations for Privacy, Data Protection and Cyber-security in the Smart Grid Environment,
- EG3: Expert Group for Regulatory Recommendations for Smart Grids Deployment,
- EG4: Expert Group for Smart Grid Infrastructure Deployment.

2.4.3 Agency for the Cooperation of Energy Regulators (ACER)

ACER Mission and Objectives[50]

ACER's missions and tasks are defined by the Directives and Regulations of the Third Energy Package, especially Regulation (EC) 713/2009 establishing the Agency.[51] In 2011, ACER received additional tasks under Regulation (EU) No 1227/2011 on wholesale energy market integrity and transparency (REMIT)[52] and in 2013 under Regulation (EU) No 347/2013 on guidelines for trans-European energy infrastructure.[53]

The overall mission of ACER as stated in its founding regulation is to complement and coordinate the work of national energy regulators at EU level and work towards the completion of the single EU energy market for electricity and natural gas.

ACER plays a central role in the development of EU-wide network and market rules with a view to enhance competition. It coordinates regional and cross-regional initiatives which favour market integration. It monitors the work of European networks of transmission

---

[48] Cf. http://ec.europa.eu/energy/en/topics/markets-and-consumers/smart-grids-and-meters/smartgrids-task-force.

[49] Cf. *supra*, at 2.3.1.2.

[50] Cf. http://www.acer.europa.eu/The_agency/Mission_and_Objectives/Pages/default.aspx.

[51] Cf. *supra*, note 23 [VP & DK].

[52] Regulation (EU) No 1227/2011 of the European Parliament and of the Council of 25 October 2011 on wholesale energy market integrity and transparency, OJ L 326, 8.12.2011, pp. 1–16 [VP & DK].

[53] Regulation (EU) No 347/2013 of the European Parliament and of the Council of 17 April 2013 on guidelines for trans-European energy infrastructure and repealing Decision No 1364/2006/EC and amending Regulations (EC) No 713/2009, (EC) No 714/2009 and (EC) No 715/2009, OJ L 115, 25.4.2013, pp. 39–75 [VP & DK].

system operators (ENTSOs) and notably their EU-wide network development plans. Finally, it monitors the functioning of gas and electricity markets in general, and of wholesale energy trading in particular.

2.4.4 National Regulatory Authorities in the EU/EEA

EU/EEA Member State[a]	Authority	Website
Austria	E-control Energie-Control Austria	www.e-control.at
Belgium	CREG Commission pour la Régulation de l'Electricité et du Gaz	www.creg.be
Bulgaria	SEWRC комисия за енергийно и водно регулиране—State Energy and Water Regulatory Commission	www.dker.bg
Croatia	HERA Hrvatska energetska regulatorna agencija—Croatian Energy Regulatory Agency	www.hera.hr
Cyprus	CERA Ρυθμιστική Αρχή Ενέργειας Κύπρου— Cyprus Energy Regulatory Authority	www.cera.org.cy
Czech Republic	ERÚ—ERO Energetický Regulační Úřad—Energy Regulatory Office	www.eru.cz
Denmark	DERA Energitilsynet—Danish Energy Regulatory Authority	www.dera.dk
Estonia	ECA Konkurentsiamet—Estonian Competition Authority—Energy Regulatory Dept	www.konkurentsiamet.ee
Finland	EMV Energiamarkkinavirasto—The Energy Market Authority	www.energiamarkkinavirasto.fi
France	CRE Commission de Régulation de l'Energie	www.cre.fr
Germany	Bundesnetzagentur—BnetzA Federal Network Agency for Electricity, Gas, Telecommunications, Posts and Railway	www.bundesnetzagentur.de

(continued)

EU/EEA Member State[a]	Authority	Website
Greece	PAE—RAE Ρυθμιστική Αρχή Ενέργειας—Regulatory Authority for Energy	www.rae.gr
Hungary	MEH—HEO Magyar Energia Hivatal—Hungarian Energy Office	www.eh.gov.hu
Iceland	OS Orkustofnun—National Energy Authority	www.os.is
Ireland	CER Commission for Energy Regulation	www.cer.ie
Italy	AEEG Autorità per l'Energia Elettrica e il Gas	www.autorita.energia.it
Latvia	PUC Sabiedrisko pakalpojumu regulēšanas komisija—Public Utilities Commission	www.sprk.gov.lv
Lithuania	NCC Valstybinė kainų ir energetikos kontrolės komisija—National Control Commission for Prices and Energy	www.regula.lt
Luxemburg	ILR Institut Luxembourgeois de Régulation	www.ilr.lu
Malta	MRA Malta Resources Authority	www.mra.org.mt
The Netherlands	ACM Autoriteit Consument & Markt—The Netherlands Authority for Consumers and Markets	www.acm.nl
Norway	NVE Norges vassdrags- og energidirektorat—Norwegian Water Resources and Energy Directorate	www.nve.no
Poland	URE—ERO Urząd Regulacji Energetyki—The Energy Regulatory Office of Poland	www.ure.gov.pl
Portugal	ERSE Entidade Reguladora dos Serviços Energéticos—Energy Services Regulatory Authority	www.erse.pt
Romania	ANRE Autoritatea Nationala de Reglementare in Domeniul Energiei	www.anre.ro

(continued)

EU/EEA Member State[a]	Authority	Website
Slovakia	URSO—RONI Úrad pre reguláciu sieťových odvetví— Regulatory Office for Network Industries	www.urso.gov.sk
Slovenia	AGEN—RS Javna Agencija Republike Slovenije za energijo—Energy Agency of the Republic of Slovenia	www.agen-rs.si
Spain	CNE La Comisión Nacional de Energía— National Energy Commission	www.cne.es
Sweden	EI Energimarknadsinpektionen—Energy Markets Inspectorate	www.ei.se
United Kingdom	Ofgem Office of Gas and Electricity Markets	www.ofgem.gov.uk

[a] Cf. http://www.entsog.eu/national-regulatory-authorities-nras

2.4.5 Selected European Organizations and Associations of Industry

	Name	Website
CEDEC	European Federation of Local Energy Companies	http://www.cedec.com
CEER	Council of European Energy Regulators	http://www.ceer.eu
EDSO	European Distribution System Operators	https://www.edsoforsmartgrids.eu
ENCS	European Network for Cyber Security	http://www.encs.eu
ENTSO-E	European Network of Transmission System Operators for Electricity	https://www.entsoe.eu
ENTSOG	European Network of Transmission System Operators for Gas	http://www.entsog.eu
ESMIG	European Smart Metering Industry Group	http://www.esmig.eu
EURELECTRIC	Union of the Electricity Industry	http://www.eurelectric.org
EUTC	European Utilities Telecom Council	http://www.eutc.org
GEODE	Verband der unabhängigen Strom- und Gasverteilerunternehmen	http://www.geode-eu.org
SEDC	Smart Energy Demand Coalition	http://sedc-coalition.eu

2.4.6 Selected European Standardization Bodies

	Name	Website
CENELEC	European Committee for Electro-technical Standardisation	http://www.cenelec.eu
ETSI	European Telecommunications Standards Institute	http://www.etsi.org

2.5 Legal Framework for Personal Data Protection in the EU

2.5.1 Context and Background of the Data Protection Law in Europe

Raymond Wacks, *Privacy. A very short introduction* **(2010)** [8]

At the most general level, the idea of privacy embraces the desire to be left alone, free to be ourselves—uninhibited and unconstrained by the prying of others. This extends beyond snooping and unsolicited publicity to intrusions upon the 'space' we need to make intimate, personal decisions without the intrusion of the state. [...]

In any event, it is clear that at the core of our concern to protect privacy lies a conception of the individual's relationship with society. Once we acknowledge a separation between the public and the private domain, we assume a community in which not only does such a division makes sense, but also institutional structure that makes possible an account of this sort.

Louis D. Brandeis and Samuel Warren, *The Right to Privacy* **(1890)** [9]

The intensity and complexity of life, attendant upon advancing civilization, have rendered necessary some retreat from the world, and man, under the refining influence of culture, has become more sensitive to publicity, so that solitude and privacy have become more essential to the individual; but modern enterprise and invention have, through invasions upon his privacy, subjected him to mental pain and distress, far greater than could be inflicted by mere bodily injury.

The notion of "privacy" is conceptualized differently in different disciplines and cultures—ranging from sociology and anthropology to applied ethics and computer science—but from the legal perspective, in Western democratic legal cultures, "privacy" is considered a fundamental right and safeguarded on multiple levels.[54] In particular:

---

[54] Further reading on privacy and data protection include, among others, [8, 10].

- At the **international** level, the right to privacy is protected by Art 12 of the Universal Declaration of Human Rights (1948);[55] however, non-binding yet standard setting, and Art 17 the International Covenant on Civil and Political Rights (1966),[56] which, on the contrary, is legally binding;
- At the **regional** level, in Europe, the system of legal protection is based on the Art 8 of the European Convention on Human Rights (ECHR)[57] and certain sector-specific instruments, namely the Convention for the Protection of Individuals with regard to Automatic Processing of Personal Data (No. 108)[58] supplemented by an additional protocol concerning supervisory authorities (No. 181).[59] The ECHR establishes the European Court of Human Rights (ECtHR)[60] in Strasbourg that hears complaints from individuals against alleged violations of the ECHR by states;
- At the **supranational** level, in the European Union, the protection is based on its Treaties,[61] the Charter of the Fundamental Rights (CFR)[62] and the secondary legislation, namely the Directives. Art 16 TFEU and Art 39 TEU both recognize the right to data protection. Art 7 of the Charter provides the right to respect for private and family life and its Art 8 provides for the protection of personal data. The Court of Justice of the EU (colloquially: ECJ)[63] in Luxembourg ensures the uniform application of the EU law;
- At the **national** level, virtually all the Western democratic states protect these two rights in their national constitutions.

The EU constitutional provisions in the Treaties are further specified in the secondary legislation, currently consisting of five main legal instruments:

- 1995 Data Protection Directive,[64]

---

[55] Cf. http://www.un.org/en/documents/udhr/.

[56] Cf. http://www.ohchr.org/en/professionalinterest/pages/ccpr.aspx.

[57] Convention for the Protection of Human Rights and Fundamental Freedoms, as amended by Protocols No. 11 and No. 14, Rome, 4 November 1950, ETS No. 5. http://conventions.coe.int/treaty/en/treaties/html/005.htm.

[58] Convention for the Protection of Individuals with regard to Automatic Processing of Personal Data, Strasbourg, 28 January 1981, ETS 181. http://www.conventions.coe.int/Treaty/en/Treaties/Html/108.htm.

[59] Additional Protocol to the Convention for the Protection of Individuals with regard to Automatic Processing of Personal Data regarding supervisory authorities and transborder data flows, Strasbourg, 8 November 2001, ETS 181. http://www.conventions.coe.int/Treaty/en/Treaties/Html/181.htm.

[60] Cf. http://www.echr.coe.int.

[61] Cf. *supra*, note 7.

[62] Charter of Fundamental Rights of the European Union, OJ C 326, 26.10.2012, pp. 391–407.

[63] Cf. http://curia.europa.eu.

[64] Directive 95/46/EC of the European Parliament and of the Council of 24 October 1995 on the protection of individuals with regard to the processing of personal data and on the free movement of such data, OJ L 281, 23.11.1995, pp. 31–50.

- 2002 ePrivacy Directive,[65] as amended by Directives: 2006/24/EC and 2009/136/EC,
- 2006 Data Retention Directive,[66]
- 2008 Data Protection Framework Decision,[67]
- Regulation 45/2001.[68]

Exhibit 2: Legal framework for privacy and personal data protection worldwide

Universal Declaration of Human Rights 1948 (UDHR)

Art 12
No one shall be subjected to arbitrary interference with his privacy, family, home or correspondence, nor to attacks upon his honour and reputation. Everyone has the right to the protection of the law against such interference or attacks.

Art 29(2)
In the exercise of his rights and freedoms, everyone shall be subject only to such limitations as are determined by law solely for the purpose of securing due recognition and respect for the rights and freedoms of others and of meeting the just requirements of morality, public order and the general welfare in a democratic society.

International Covenant on Civil and Political Rights 1966 (ICCPR)

Art 17
1. No one shall be subjected to arbitrary or unlawful interference with his privacy, family, home or correspondence, nor to unlawful attacks on his honour and reputation.
2. Everyone has the right to the protection of the law against such interference or attacks.

European Convention on Human Rights 1950 (ECHR)

Art 8
1. Everyone has the right to respect for his private and family life, his home and his correspondence.

(continued)

---

[65] Directive 2002/58/EC of the European Parliament and of the Council of 12 July 2002 concerning the processing of personal data and the protection of privacy in the electronic communications sector (Directive on privacy and electronic communications), OJ L 201, 31.7.2002, pp. 37–47.

[66] Directive 2006/24/EC of the European Parliament and of the Council of 15 March 2006 on the retention of data generated or processed in connection with the provision of publicly available electronic communication services or of public communications networks and amending Directive 2002/58/EC, OJ L 105, 13.4.2006, pp. 54–63. The Data Retention Directive has been recently invalidated by the Court of Justice of the European Union, in joint cases C-293/12 and C-594/12, as entailing a "wide-ranging and particularly serious interference with the fundamental rights to respect for private life and to the protection of personal data, without that interference being limited to what is strictly necessary" [11].

[67] Council Framework Decision 2008/977/JHA of 27 November 2008 on the protection of personal data processed in the framework of police and judicial cooperation in criminal matters, OJ L 350, 30.12.2008, pp. 60–71.

[68] Regulation 45/2001 of the European Parliament and of the Council of 18 December 2000 on the protection of individuals with regard to the processing of personal data by the Community institutions and bodies and on the free movement of such data, OJ L 8, 12.1.2001, pp. 1–22.

2. There shall be no interference by a public authority with the exercise of this right except such as is in accordance with the law and is necessary in a democratic society in the interests of national security, public safety or the economic well-being of the country, for the prevention of disorder or crime, for the protection of health or morals, or for the protection of the rights and freedoms of others.

Charter of Fundamental Rights of the European Union 2001 (CFR EU)

Art 7—Respect for private and family life
Everyone has the right to respect for his or her private and family life, home and communications.

Art 8—Protection of personal data
1. Everyone has the right to the protection of personal data concerning him or her.
2. Such data must be processed fairly for specified purposes and on the basis of the consent of the person concerned or some other legitimate basis laid down by law. Everyone has the right of access to data which has been collected concerning him or her, and the right to have it rectified.
3. Compliance with these rules shall be subject to control by an independent authority.

Art 51(1)—Field of application
The provisions of this Charter are addressed to the institutions, bodies, offices and agencies of the Union with due regard for the principle of subsidiarity and to the Member States only when they are implementing Union law (…)

Art 52(1)—Scope and interpretation of rights and principles
Any limitation on the exercise of the rights and freedoms recognised by this Charter must be provided for by law and respect the essence of those rights and freedoms. Subject to the principle of proportionality, limitations may be made only if they are necessary and genuinely meet objectives of general interest recognised by the Union or the need to protect the rights and freedoms of others.

Treaty on the Functioning of the European Union (TFEU)

Art 16 (ex Art 286)
1. Everyone has the right to the protection of personal data concerning them.
2. The European Parliament and the Council, acting in accordance with the ordinary legislative procedure, shall lay down the rules relating to the protection of individuals with regard to the processing of personal data by Union institutions, bodies, offices and agencies, and by the Member States when carrying out activities which fall within the scope of Union law, and the rules relating to the free movement of such data. Compliance with these rules shall be subject to the control of independent authorities.
The rules adopted on the basis of this Article shall be without prejudice to the specific rules laid down in Article 39 of the Treaty on European Union.

Treaty on the European Union (TEU)

Art 39
In accordance with Article 16 of the Treaty on the Functioning of the European Union and by way of derogation from paragraph 2 thereof, the Council shall adopt a decision laying down the rules relating to the protection of individuals with regard to the processing of personal data by the Member States when carrying out activities which fall within the scope of this Chapter (Specific Provisions on the Common Foreign and Security Policy), and the rules relating to the free movement of such data. Compliance with these rules shall be subject to the control of independent authorities.

2.5.2 Basic Data Protection Terminology

1995 Data Protection Directive

Article 2
Definitions

For the purposes of this Directive:

(a) 'personal data' shall mean any information relating to an identified or identifiable natural person ('data subject'); an identifiable person is one who can be identified, directly or indirectly, in particular by reference to an identification number or to one or more factors specific to his physical, physiological, mental, economic, cultural or social identity;

(b) 'processing of personal data' ('processing') shall mean any operation or set of operations which is performed upon personal data, whether or not by automatic means, such as collection, recording, organization, storage, adaptation or alteration, retrieval, consultation, use, disclosure by transmission, dissemination or otherwise making available, alignment or combination, blocking, erasure or destruction;

[...]

(d) 'controller' shall mean the natural or legal person, public authority, agency or any other body which alone or jointly with others determines the purposes and means of the processing of personal data; where the purposes and means of processing are determined by national or Community laws or regulations, the controller or the specific criteria for his nomination may be designated by national or Community law;

(e) 'processor' shall mean a natural or legal person, public authority, agency or any other body which processes personal data on behalf of the controller;

[...]

(h) 'the data subject's consent' shall mean any freely given specific and informed indication of his wishes by which the data subject signifies his agreement to personal data relating to him being processed.

2.5.3 Data Protection Principles

2.5.3.1 General Data Protection Principles

The EU data protection framework establishes a **"five-layer"** set of principles for processing personal data. **First layer** contains a number of general principles applicable to the processing of any personal data. Accordingly, these data must be:

- **fairly and lawfully** processed—Art 6(1)(a) of the 1995 Data Protection Directive

- **data minimization**

  - collected for specific, explicitly defined and legitimate purposes—Art 6(1)(b)
  - not further processed in a way incompatible with those—Art 6(1)(b)
  - retained only for as long as is necessary to fulfil that purpose—Art 6(1)(c) (implicitly)

- **data quality**

  - adequate, relevant, and not excessive in relation to the purposes for which they are collected and/or further processed—Art 6(1)(c)
  - accurate and, where necessary, kept up to date—Art 6(1)(d)

- based on one of the **legitimate basis** for processing—Art 7

  - unambiguous consent of the data subject
  - performance of a contract to which the data subject is a party
  - compliance with a legal obligation of the data controller
  - protection of the vital interest of the data subject
  - performance of the task carried out in the public interest or exercise of official authority
  - legitimate interest pursued by the controller

- **data anonymization**—Art 6(1)(e)

  - kept in a form which permits identification of data subjects for no longer than is necessary for the purposes for which the data were collected or for which they are further processed

- **processed confidentially**—i.e. "any person acting under the authority of the controller or of the processor, including the processor himself, who has access to personal data must not process them except on instructions from the controller, unless he is required to do so by law"—Art 16
- **processed securely**—i.e. required are appropriate technical and organizational measures to protect personal data against accidental or unlawful destruction or accidental loss, alteration, unauthorized disclosure or access, in particular where the processing involves the transmission of data over a network, and against all other unlawful forms of processing—Art 17
- **notified to a relevant supervisory authority**—i.e. controller must notify the national supervisory authority before carrying out any wholly or partly automatic processing operation—Art 18(1); subject to certain exceptions, e.g. appointing the in-house data protection official (Art 18(2)).

2.5.3.2 Processing of Sensitive Data

On top of that, the 1995 Data Protection Directive introduces a stricter and prohibitive regime for sensitive data. Therefore, as the **second layer**, **processing of certain categories of data** is prohibited unless special safeguards, listed below, are met.

These data include those revealing racial or ethnic origin, political opinions, religious or philosophical beliefs, trade-union membership, and the processing of data concerning health or sex life (Art 8(1)).

Art 8(2) lists exceptions from this provision (safeguards), that is:

- explicit consent of the data subject,
- obligations in the field of employment law,
- protection of the vital interest of the data subject where the data subject is physically or legally incapable of giving his consent,
- legitimate activities of a foundation, association or any other non-profit body,
- when personal data were manifestly made public by data subject,
- purposes of preventive medicine, medical diagnosis, care or other treatment— Art 8(3).

Member States may, for reasons of substantial public interest, lay down exemptions in addition to the above-mentioned (Art 8(4)). Processing of data relating to offences, criminal convictions, or security measures may be carried out only under the control of official authority (Art 8(5)). Processing of personal data carried out solely for journalistic purposes or the purpose of artistic or literary expression is allowed (Art 9).

2.5.3.3 Transfer of Personal Data Outside the European Economic Area

As a **third layer**, transfer of personal data to countries outside the European Economic Area (EEA)[69] without adequate level of protection is prohibited (Art 25), unless it is covered by one of the following exceptions—Art 26(1):

- explicit unambiguous consent of the data subject
- contract or precontractual measures
- contract between controller and a third party in the interest of the data subject
- important public interest
- vital interest of the data subject
- transfer from a public register
- authorization by Member State—Art 26(2)

It is the European Commission that determines what jurisdictions provide the adequate level of protection (Art 25(6)).[70]

---

[69] Due to a relatively complex nature of European integration, the European Economic Area (EEA) consists of the European Union (i.e. 28 Member States) as well as Norway, Iceland, and Liechtenstein and provides for a free movement of goods, persons, services, and capital between the contracting parties. Switzerland, on the contrary, maintains a bilateral relationship with the EU/EEA.

[70] The current list can be found at http://ec.europa.eu/justice/policies/privacy/thridcountries/index_en.htm.

2.5.3.4 Sector-Specific Rules

The **fourth layer** consists of sector-specific rules, laid down by the ePrivacy and Data Retention directives as well as by self-regulation. The ePrivacy Directive regulates the processing of personal data and the protection of privacy in the electronic communications sector, namely the traffic and location data. The Data Retention Directive, recently invalidated,[71] regulated the data retention period regarding publicly available communications services and public communications networks.[72] When it comes to self-regulation, the Data Protection Directive allows for the creation of *"codes of conduct intended to contribute to the proper implementation of* (data protection framework), *taking account of the specific features of the various sectors"* (Art 27(1)).

2.5.3.5 Specific Legal Relationship Between the Data Subject and Data Controller

The **fifth layer** deals with specific legal relationship between the data subject and data controller, e.g. established by a contract.

2.5.4 The Rights of the Individuals with Regard to Processing Their Personal Data

The data subject has the following rights regarding processing her personal data:

- the right to be informed about processing his/her personal data in a clear and understandable language—Art 12(a),
- the right to access to own personal data—Art 12(a),
- the right to rectify any wrong or incomplete information—Art 12(b),
- the right, in some cases, to object the processing on legitimate grounds—Art 14,
- the right not to be subject to an automated decision intended to evaluate certain personal aspects relating to the data subject as his performance at work, creditworthiness, reliability, conduct—Art 15,
- the right to judicial remedy and to receive compensation from the data controller for any damage suffered (short of *vis maior*)—Arts 23–23, respectively.

---

[71] Cf. *supra*, note 68.

[72] Despite invalidation in April 2014 (cf. *supra*, note 68), the Data Retention Directive is still mentioned here as national laws enacted in implementation of that Directive would for the time being remain in force, unless retracted or invalidated by national higher courts.

These data subject's rights correspond to the data controller's obligations to:

- ensure the data subject's rights are duly observed,
- ensure observance of the data minimization principle,
- ensure observance of the criteria for making the data processing legitimate (e.g. consent or performance of the contract),
- safeguard confidentially of processing,
- safeguard security of processing,
- notify processing of personal data to the national data protection authority (DPA),
- in case of the transfer to the third countries—ensure if these countries provide adequate level of protection (in general).

2.5.5 The Reform of the EU Data Protection Framework

Since January 2012, the EU data protection framework is undergoing a substantial reform process.[73] The proposed new framework is believed to:[74]

strengthen online privacy rights and boost Europe's digital economy. Technological progress and globalisation have profoundly changed the way our data is collected, accessed and used. [...] A single law will do away with the current fragmentation and costly administrative burdens [...]. The initiative will help reinforce consumer confidence in online services, providing a much needed boost to growth, jobs and innovation in Europe. [...]

Key changes in the reform include:

- A single set of rules on data protection, valid across the EU. Unnecessary administrative requirements, such as notification requirements for companies, will be removed. [...]
- Instead of the current obligation of all companies to notify all data protection activities to data protection supervisors [...], the Regulation provides for increased responsibility and accountability for those processing personal data.
- [...] companies and organisations must notify the national supervisory authority of serious data breaches as soon as possible (if feasible within 24 h).

---

[73] European Commission, Proposal for a Regulation of the European Parliament and of the Council on the protection of individuals with regard to the processing of personal data and on the free movement of such data (General Data Protection Regulation), Brussels, 25 January 2012, COM (2012)11 final; *hereinafter*: the GDPR *or* the EU General Data Protection Regulation.

[74] European Commission, *Commission proposes a comprehensive reform of data protection rules to increase users' control of their data and to cut costs for businesses*, press release, IP/12/46, Brussels, 25 January 2012. http://europa.eu/rapid/press-release_IP-12-46_en.htm.

- Organisations will only have to deal with a single national data protection authority in the EU country where they have their main establishment. Likewise, people can refer to the data protection authority in their country, even when their data is processed by a company based outside the EU. Wherever consent is required for data to be processed, it is clarified that it has to be given explicitly, rather than assumed.
- People will have easier access to their own data and be able to transfer personal data from one service provider to another more easily (right to data portability). […]
- A 'right to be forgotten' will help people better manage data protection risks online: people will be able to delete their data if there are no legitimate grounds for retaining it.
- EU rules must apply if personal data is handled abroad by companies that are active in the EU market and offer their services to EU citizens.
- Independent national data protection authorities will be strengthened so they can better enforce the EU rules at home. They will be empowered to fine companies that violate EU data protection rules. This can lead to penalties of up to €1 million or up to 2 % of the global annual turnover of a company. […]

In the spirit of the accountability principle, the proposed framework introduces a requirement to obligatorily conduct a data protection impact assessment (DPIA) for specific processing operations. As it will be demonstrated *infra*, at 2.7, a DPIA constitutes a main approach to achieve personal data protection goals in smart grid and smart metering systems in the EU.

Proposal for a Regulation of the European Parliament and of the Council on the protection of individuals with regard to the processing of personal data and on the free movement of such data (General Data Protection Regulation) (2012)

Article 33
Data protection impact assessment

1. Where processing operations present specific risks to the rights and freedoms of data subjects by virtue of their nature, their scope or their purposes, the controller or the processor acting on the controller's behalf shall carry out an assessment of the impact of the envisaged processing operations on the protection of personal data.

2. The following processing operations in particular present specific risks referred to in paragraph 1:

(a) a systematic and extensive evaluation of personal aspects relating to a natural person or for analysing or predicting in particular the natural person's economic situation, location, health, personal preferences, reliability or behaviour, which is based on automated processing and on which measures are based that produce legal effects concerning the individual or significantly affect the individual;
(b) information on sex life, health, race and ethnic origin or for the provision of health care, epidemiological researches, or surveys of mental or infectious diseases, where the data are processed for taking measures or decisions regarding specific individuals on a large scale;

(c) monitoring publicly accessible areas, especially when using optic-electronic devices (video surveillance) on a large scale;
(d) personal data in large scale filing systems on children, genetic data or biometric data;
(e) other processing operations for which the consultation of the supervisory authority is required […].

3. The assessment shall contain at least a general description of the envisaged processing operations, an assessment of the risks to the rights and freedoms of data subjects, the measures envisaged to address the risks, safeguards, security measures and mechanisms to ensure the protection of personal data and to demonstrate compliance with this Regulation, taking into account the rights and legitimate interests of data subjects and other persons concerned.

4. The controller shall seek the views of data subjects or their representatives on the intended processing, without prejudice to the protection of commercial or public interests or the security of the processing operations.

At the time of writing (November 2014), the proposal is still pending in the European Parliament and the Council. It is estimated that the legislation will be adopted in early 2015 with the entry into force in 2016 or 2017.

2.6 The Interaction of Smart Grid and Smart Metering Systems with Data Protection Law

2.6.1 Application of the Data Protection Law for Smart Grid and Smart Metering Systems

The EU data protection framework places substantial limitations to personal data processing. Because smart grid and smart metering systems unavoidably process personal data as part of their routine technical processes, and indeed derive added value from such processing, they need to be examined under this light, at least as far as their EU application is concerned.

Data protection (information privacy) issues are evidently raised in jurisdictions outside the EU as well: the US institutions have already identified the privacy concerns caused by the related personal data processing [12]. However, the analysis that follows will focus only on the EU data protection model, because of the formal processing requirements imposed by it as well as extensive work already undertaken at EU level to create a comprehensive regulatory environment that is detailed enough in order to assist other jurisdictions. These pertain to:

(a) the role of its actors,
(b) their obligations,
(c) monitoring and oversight, and
(d) individual rights and remedies.

It is around the above main personal data processing issues that the analysis that follows shall develop. The authors believe that any smart grid implementation within a jurisdiction where information privacy rights are acknowledged in one way or another will inevitably have to address at least the above central issues. The EU paradigm, although deriving from formal and strict data protection rules that are met only in a handful of jurisdictions outside its borders, is considered useful in identifying potential challenges and providing at least a reasoning how best to address them while balancing industry interests and individual rights.

In attempting the above, a number of preliminary data protection issues shall be considered resolved: this relates mostly to the fact that data protection legislation is applicable because personal data are being processed within smart grid implementations.[75] This should be perceived as referring to anonymized (or pseudonymous) data as well, given that it is normally technically possible to track the data back to their source. Other technical matters referring to the categories and type of data collected are of no concern to this analysis, given general data protection law application (however, the issue of deriving sensitive data from such processing is addressed below). The same applies to more specialized issues, for instance, whether electronic communication regulation is also applicable or whether trans-border data flows need to be regulated, given that such issues need to be addressed on an *ad hoc* basis. Instead, the analysis that follows aims at identifying only the broader data protection issues that smart grid implementations face or are bound to face and that will probably need to be resolved since the first steps of their implementation.

2.6.2 Distinction Between Data Processors and Data Controllers

The actors involved in the smart grid chain of service are numerous—ranging from transmission system operators (TSO) to distribution system operators (DSO) to metering operators—and their roles may vary substantially depending on the actual implementation details. However, assigning roles within the smart grid system may prove important with regard to application of data protection provisions. The 1995 Data Protection Directive distinguishes between controllers and processors, essentially placing only the former under its scrutiny (cf. *supra*, at 2.5.2). Accordingly, only controllers are liable to apply the law to their processing.[76] It is therefore important in view of adequate application of data protection legislation to be able to distinguish among the different actors.

---

[75] See Article 2 of the EU Data Protection Directive, Article 29 Data Protection Working Party, Opinion 12/2011 on smart metering (4 April 2011), p. 7, cf. further [13].
[76] Art 6.2.

Identification of the problem in this case only serves to demonstrate its complexity. Even within established processing circumstances, the distinction between controller and processor is challenged and at times needs to be resolved by courts.[77] Given the number of participants involved in a typical smart grid system, it is likely that their own division of roles and obligations, to be described in the relevant contractual framework, will not remain unchallenged or untested both by local data protection authorities (DPAs) and the individuals concerned. The fact that implementations may vary among different states does not assist harmonization or at least a coordinated approach.

The Art 29 Working Party, an independent EU advisory body on personal data protection, attempted to address this issue by reference to a simple smart grid model composed of energy supplier(s), network operator(s) and "other parties"—the latter to denote the multitude of *"parties who could potentially be processing personal data in the course of fulfilling their role in a programme to implement smart meters"*. In this context, while energy suppliers are expected to constitute controllers, this might (but equally might not) be the case with network operators. On the other hand, *"other parties"* are expectedly beyond categorization and are therefore only reminded of the basic data protection principles to this end, namely that any decision-making or any self-serving personal data processing means that they are automatically excluded from their (evidently preferred) processor status. The importance of role separation is also acknowledged in the 2012 Recommendation, whereby however, Member States are more or less left alone when being required to *"clearly determine the roles and responsibilities of data controllers and data processors. They should be compatible with their respective obligations set out in Directive 95/46/EC* (the 1995 Data Protection Directive)".[78]

While it appears that a harmonized approach for all EU Member States with regard to role separation cannot be adopted, at least at this stage of smart grid systems deployment, and the texts and instruments released to-date may serve as guidance at best, it is important for these systems to take into account even since their drafting stage the relevant legal requirements. Local data protection authorities could provide valuable guidance to this end—and, because it will also be the same that will accept individual complaints in the future as well as process the relevant files even in the event of court referral, it is recommended for the smart grid industry to act proactively in this matter.

---

[77] This was, for instance, the case with search engines, whereby, although relevant operators apparently considered themselves as processors for data protection purposes, it was only in early 2014 that the Court of Justice of the European Union clarified that they too need to be considered data controllers (cf. Court of Justice of the European Union, *Google Spain SL and Google Inc. v Agencia Española de Protección de Datos (AEPD) and Mario Costeja González*, C-131/12).

[78] Cf. *supra*, note 41, par. 21.

2.6.3 The Purpose Limitation Principle

Once the roles within the smart grid context haven been established, data controllers
need to ensure the adequate application of the data protection rules and principles
within their personal data processing. This might prove a far from straightforward
task that, again, cannot be easily streamlined even within the EU, despite of the fact
that the same legislation (i.e. the 1995 Data Protection Directive) applies. This is
due, first, to diverse Member State approaches to the data protection rules and,
second, to the different smart grid implementations adopted within these countries.
The former category of difficulties will perhaps be resolved when the draft EU
General Data Protection Regulation comes into effect, although it must be noted
that smart grid systems are expected to attract specialized, secondary, regulation.[79]
However, the latter category of problems, that of diverse smart grid implementations among Member States, is not expected to be resolved in the immediate future
and in any case not until such systems have met generalized use among Member
States. For such intermediate period, regulatory guidance will probably have to take
place at an *ad hoc* basis, most likely with local DPAs at least as far as personal data
processing is concerned.

Substantial restrictions to smart grid systems are expected to be placed through
application of a basic EU data protection principle, the purpose limitation principle:
according to the relevant provisions, personal data must be "*collected for specified,
explicit and legitimate purposes and not further processed in a way incompatible
with those purposes*".[80] This is essentially a practical principle that imposes concrete obligations upon data controllers, namely to collect and process data only for
known and declared purposes—and to delete them after they have been served.
Consequently, retention of data for other, undeclared, and irrelevant purposes to
these of the original collection is prohibited. In the smart grid context, purpose-
relevant processing will probably include all stages for the provision of the relevant
services (installation, operation and monitoring of the system, including billing) and
perhaps limited marketing of the same but most likely not, for instance, processing
for security or health-related purposes. Time periods over which such data are kept
in an identifiable format are important and data controllers will probably need to
justify them in front of the competent DPA. In the event that electronic communications are used with regard to provision of the smart grid services a closer look to
the relevant provisions (in particular, whether this is a public network or not)
probably needs to take place. Finally, aggregate data processing for research or
better system management purposes may probably be executed, but only after
personal data have been made anonymous.[81]

---

[79] Cf. the analysis on impact assessments or certification, *infra* at 2.8.5 and 2.8.7, respectively.

[80] Art 6(1)(b).

[81] However, on the anonymization of personal data in the smart grid context, cf. [14].

2.6.4 Data Storage

The issue of data storage is connected to the purpose limitation principle: personal information, regardless whether stored in the subscribers' meters or kept in the operators' databases, needs to be kept for a time period that will be determined by the purposes of the relevant processing. As outlined above, such time periods may differ substantially, depending on the operations of the actual system: for instance, use and billing purposes might include a storage period of a few months, whereas security-related purposes, if ever connected and permitted under the smart grid context, would perhaps justify data retention for longer time periods.

2.6.5 Fair and Lawful Processing of Personal Information

According to the 1995 Data Protection Directive, *"personal data must be processed fairly and lawfully"*.[82] This double criterion is applicable both to the stage of data collection and to the stage of the actual data processing—in fact, until deletion of the data concerned.[83] While the fairness criterion is further connected in the text of the Directive with the right to information,[84] the lawfulness requirement will generally encompass proper adherence to all applicable legislative provisions. Consequently, smart grid systems will have to observe both the general data protection provisions applicable with regard to their processing (for the time being, national data protection acts that in the near future might be replaced by the EU General Data Protection Regulation) and provide fair chance to data subjects to be aware and alert, in order also to properly consent, with regard to such processing. In practice, this principle sets general requirements that are expected to affect both the smart grid system design (providing, for instance, information to individuals or assisting them while offering their data to the system) and the smart grid process, in the sense that data controllers will have to take into account the general data protection legislation and, for instance, take all necessary actions in front of the competent DPAs with regard to their processing.

2.6.6 The Principle of Proportionality

As per another basic EU data protection principle, personal data processing needs to be proportionate to its purposes: *"personal data must be: [...] adequate, relevant and not excessive in relation to the purposes for which they are collected and/or*

---

[82] Art (6)(1)(a).
[83] Art 2.
[84] Arts 11–12, preamble 38 as well as *infra*, at 2.6.9.

further processed".[85] While general in nature, the principle of proportionality may be used to place concrete obligations upon data controllers, namely to devise their systems in such a way so as to process as little information as possible with regard to their, already declared, purposes for such processing. In the smart grid context, this might prove limiting, in the sense that it will essentially affect the "smartness" of the system: despite the fact that from a technical point of view the more data are fed into the system the smarter the system becomes, data protection regulations place restrictions to the categories of data collected. Smart meters installed in subscribers' premises are only allowed to collect these data that are needed in order to provide the smart grid service. Other data, that could enhance the service, are not allowed to be collected (although, individual consent needs to be carefully leveraged in this context). The principle of proportionality may at times conflict with smart grid systems, that by definition will wish to expand their processing scope in order to learn and improve their service, and therefore, careful balancing of individuals' right to data protection and individuals' (and providers') wish for better services will need to take place (most likely, by DPAs and, ultimately, by the competent courts).

2.6.7 Data Quality

Little doubt exists that personal data processed by smart grid systems need to be *"accurate"* and *"kept up to date"*.[86] In fact, it is to the interest of both smart grid systems' operators and individuals that the personal data processing executed in this context makes use only of the latest and most relevant information. It is only in this way that smart grid systems shall develop their full potential, to the benefit of all parties concerned. Therefore, the basic EU data protection principle of data quality is expected to be the least problematic, or unwelcome, while implementing smart grid systems across the EU.

2.6.8 Monitoring and Oversight of Smart Grid Data Controllers

The issue of monitoring and oversight of smart grid data controllers is relatively resolved within EU data protection jurisdictions, given that the local DPA will generally undertake all relevant tasks. Given also that smart grid-related personal data processing is not normally expected to cross national borders, the designation of such competent authority will be a straightforward task. Data controllers, once

---

[85] Art (6)(1)(c).
[86] Art (6)(1)(d).

properly identified (cf. *supra*, at 2.6.2), will have to contact the DPA concerned in order to establish the necessary steps they need to undertake in view of the lawful processing of personal data within their systems: such steps, under the current EU regime, may include anything from a simple notification to prior consultation and even prior license acquisition, for instance in the event that sensitive data (e.g. health data) may be inferred from their system. It is perhaps exactly at this point where difficulties for smart grid data controllers may be found: particularly international companies may find themselves required to fulfil different obligations among different EU states, regardless of the fact that such states otherwise apply the same basic law (i.e. the 1995 Data Protection Directive). These difficulties will be resolved once the EU General Data Protection Regulation comes into effect.

However, application of the EU General Data Protection Regulation might not prove the panacea smart grid data controllers might expect: a series of new instruments are introduced in its provisions, most of which will probably find application in the smart grid context. As discussed below (cf. *infra*, at 2.8), impact assessments, certification, or even privacy by design may all become relevant in smart grid systems. These requirements, although streamlined across the EU and aimed at reducing current bureaucratic burden (mostly in the form of notification to DPAs), are expected to impose substantial obligations to smart grid data controllers, that might even, under extreme circumstances, entail redesign of already deployed systems.

2.6.9 The Scope and Exercise of the Data Subject's Rights

A crucial part of the EU data protection framework refers to the set of specific, enforceable rights afforded to data subjects in order to allow them to monitor and control the processing of their personal data. These rights pertain to the individuals' rights: (1) to be informed that their data are being collected and processed, (2) to access such data, as well as (3) to object in the event they consider that such processing is unlawful. These rights are equally applicable in the smart grid and smart metering systems that process personal data.[87]

2.6.9.1 The Right to Information

With regard to the individuals' right to information, the 1995 Data Protection Directive distinguishes between cases where the data subject herself has given the data and cases where these data have been collected by third parties (intermediaries). In the first scenario, the data controllers need to inform the individuals about —at least—the following:[88]

---

[87] Cf. *infra*, note 103.
[88] Art 11.

Member States shall provide that the controller or his representative must provide a data subject from whom data relating to himself are collected with at least the following information, except where he already has it:

(a) the identity of the controller and of his representative, if any;
(b) the purposes of the processing for which the data are intended;
(c) any further information such as

- the recipients or categories of recipients of the data,
- whether replies to the questions are obligatory or voluntary, as well as the possible consequences of failure to reply,
- the existence of the right of access to and the right to rectify the data concerning him

in so far as such further information is necessary, having regard to the specific circumstances in which the data are collected, to guarantee fair processing in respect of the data subject.

The list is identical in the case of intermediary involvement. However, in this scenario, the EU law adds that:[89]

[w]here the data have not been obtained from the data subject, Member States shall provide that the controller or his representative must at the time of undertaking the recording of personal data or if a disclosure to a third party is envisaged, no later than the time when the data are first disclosed provide the data subject with [information as listed in Art 10].

Both scenarios (i.e. direct and indirect personal data collection) are likely to be of relevance in the smart grid and smart metering systems context. As far as direct collection and processing is concerned, the list provided in Art 11 sets the minimum information that needs to be given directly from the data controller to the data subject. Supposedly, once a contract for the provision of services is agreed between them, this information may be included in one or more of its terms.

Things are expected to be less straightforward when information has not been obtained directly from the data subject, for instance when personal data are exchanged between controllers that provide additional services or are necessary to the operation of a particular smart grid system. In this case, the right to information is triggered as described above; flexibility is warranted through application of the principle of proportionality in this case. (Member State national implementations may differ in this regard and should therefore be closely examined.) The ultimate judge, when and how such principle of proportionality applies, will evidently be a DPA concerned. If data controllers find no common ground with it, competent national courts are expected to be the final recourse for all parties concerned (meaning that individuals could also challenge for themselves already reached decisions by their local DPA and the smart grid industry).

---

[89] Art 11(1).

2.6.9.2 The Right to Access Information

The EU law does not only afford individuals with the right to know that their data are being processed: they also can access copies of such information. The right to access is a central piece of the EU data protection framework—much more so given that data subjects frequently are not informed that their data are being processed (e.g. because that would be a disproportionate exercise). Consequently, the right to access and get copies of data stored by data controllers (and therefore also to be able to ask whether this is indeed the case) is of the highest importance for the data protection purposes.

At any event, once asked, data controllers are obliged to provide to the data subject with the following information:[90]

> Member States shall guarantee every data subject the right to obtain from the controller:
>
> (a) without constraint at reasonable intervals and without excessive delay or expense:
>
> • confirmation as to whether or not data relating to him are being processed and information at least as to the purposes of the processing, the categories of data concerned, and the recipients or categories of recipients to whom the data are disclosed,
> • communication to him in an intelligible form of the data undergoing processing and of any available information as to their source,
> • knowledge of the logic involved in any automatic processing of data concerning him [...]

This is a right that will need to be closely observed by smart grid data controllers as well. A mechanism of access to personal data must be established and information on its particular details would best be provided by means of the relevant services contract, once entered with subscribers. For those data controllers who do not establish a direct relationship with data subjects whose data they process, an access mechanism needs to be devised in cooperation with other data controllers that maintain such relationships. The main idea in this case is that individuals (and DPAs alike, as it is them who monitor compliance with this obligation) need to be able to establish where their data are found and to also be able to obtain copies thereof. The retention of data for billing purposes could constitute a useful tool to this end (in the event that electronic communication services are involved in the smart grid process, special data retention rules might also apply).

In view also of the 1995 Data Protection Directive's additional requirement that individuals be informed of (and also be able to object to) the logic in automated decision-making systems (as smart grid systems might be characterized), data controllers will most likely benefit from the close cooperation with their local DPAs while devising their internal policies.[91]

---

[90] Art 12.

[91] Art 15.

1. Member States shall grant the right to every person not to be subject to a decision which produces legal effects concerning him or significantly affects him and which is based solely on automated processing of data intended to evaluate certain personal aspects relating to him, such as his performance at work, creditworthiness, reliability, conduct, etc.

2. Subject to the other Articles of this Directive, Member States shall provide that a person may be subjected to a decision of the kind referred to in paragraph 1 if that decision:

  (a) is taken in the course of the entering into or performance of a contract, provided the request for the entering into or the performance of the contract, lodged by the data subject, has been satisfied or that there are suitable measures to safeguard his legitimate interests, such as arrangements allowing him to put his point of view; or
  (b) is authorized by a law which also lays down measures to safeguard the data subject's legitimate interests.

Finally, it should also be noted that access to information in the case of smart grid systems could presumably entail access to subscribers' data files[92] (for financial or other considerations): such an option, although by no means replacing the legal requirements described above for smart grid data controllers, could enhance the general data protection purposes in the smart grid context.

2.6.9.3 The Right to Object

Once data subjects have been made aware of the fact that their personal data are being processed and have also acquired a copy therefrom, the 1995 Data Protection Directive allows them to object if they have a (lawful) reason to. According to Art 14(a), data subjects have a right:

[...] to object at any time on compelling legitimate grounds relating to his particular situation to the processing of data relating to him, save where otherwise provided by national legislation. Where there is a justified objection, the processing instigated by the controller may no longer involve those data.

They also have the option to check that lawful requests for rectification have been executed:[93]

---

[92] As is evidently the case in the UK, see "Energy companies agree to develop new data sharing systems", http://www.privacylaws.com/UK_enews_June14_1.

[93] Art 12.

[...] data subject [has] the right to obtain from the controller: [...]

(b) as appropriate the rectification, erasure or blocking of data the processing of which does not comply with the provisions of this Directive, in particular because of the incomplete or inaccurate nature of the data;

(c) notification to third parties to whom the data have been disclosed of any rectification, erasure or blocking carried out in compliance with (b), unless this proves impossible or involves a disproportionate effort.

The right to object therefore concludes the special protection to individuals: information and access would be incomplete without the persons concerned being able to react on the basis of their findings.

In the smart grid context, requests for rectification may be based on different grounds and circumstances, ranging from basic applications to rectify incorrect personal data kept in data controllers' systems to disputes on data controllers' right to process information altogether. Because the right to object is essentially a practical right, being based upon the adequate exercise and findings of the rights to information and access that precede it, it is difficult to foresee its actual practice by the subscribers concerned. Smart grid data controllers evidently need to be able to demonstrate compliance with lawful requests: on the other hand, they must be equally prepared to dispute, both in front of their competent data protection authorities and, if needed, courts, claims that are not based on law or that threaten to disproportionately limit the processing capabilities of their, by definition, "smart" systems.

2.6.10 Legal Basis for Processing—Subscribers' Consent

The processing of personal data not only has to comply with the principles included in this chapter but also has to be justified under Art 7 of the 1995 Data Protection Directive.

Article 7
[Criteria for making data processing legitimate]

Member States shall provide that personal data may be processed only if:

(a) the data subject has unambiguously given his consent; or
(b) processing is necessary for the performance of a contract to which the data subject is party or in order to take steps at the request of the data subject prior to entering into a contract; or
(c) processing is necessary for compliance with a legal obligation to which the controller is subject; or
(d) processing is necessary in order to protect the vital interests of the data subject; or

(e) processing is necessary for the performance of a task carried out in the public interest or in the exercise of official authority vested in the controller or in a third party to whom the data are disclosed; or

(f) processing is necessary for the purposes of the legitimate interests pursued by the controller or by the third party or parties to whom the data are disclosed, except where such interests are overridden by the interests for fundamental rights and freedoms of the data subject […].

While not a data protection principle itself, individual consent is a basic EU data protection law notion. Such consent needs to be a *"freely given specific and informed indication of [the individual's] wishes by which the data subject signifies his agreement to personal data relating to him being processed."*[94] However, in the smart grid context, it might prove unsuitable, mostly due to the fact that the individuals concerned can freely withdraw it. This is why the legal basis of performance of a contract might be of more relevance to the smart grid purposes: once an individual applies for or has entered a contract for the provision of smart grid services, this will perhaps constitute a more relevant legal basis governing the same individual's personal data processing (also meaning that the relevant contractual provisions shall apply to the same end).[95] At any event, it should be noted that establishment of a legal basis does not in any case mean that the basic data protection principles[96] cease to apply—on the contrary, in order for personal data processing to be lawful both a legal basis and the general requirements set by these principles ought to be met. In the smart grid context, individual consent means that all information pertaining to the processing needs to be made available to the individual before entering the relevant subscription contract (from which point the contractual terms shall apply). In addition, entering such contract must be indeed *"free"*, meaning for instance that sweeping initiatives whereby large portions of the population shall have no other option than to enter a smart grid system may only be entered after prior consultation with the competent DPAs, unless providers wish to offer individual with a way to refuse and meaningful alternatives to continue receiving the relevant services.

2.6.11 Security and Confidentiality of Data Processing

The issue of security is central both to the data protection and to the smart grid purposes. With regard to data protection, it is expressly set in the 1995 Data Protection Directive that:

---

[94] Art 2.

[95] Cf. *infra*, note 103, pp. 11–12.

[96] Art 6.

> Article 17
> **Security of processing**
>
> 1. Member States shall provide that the controller must implement appropriate technical and organizational measures to protect personal data against accidental or unlawful destruction or accidental loss, alteration, unauthorized disclosure or access, in particular where the processing involves the transmission of data over a network, and against all other unlawful forms of processing.
>
> Having regard to the state of the art and the cost of their implementation, such measures shall ensure a level of security appropriate to the risks represented by the processing and the nature of the data to be protected.

Accordingly, data controllers must choose their data processors carefully and enter written contracts with them.[97] In addition, personal data breaches need to be notified to the data protection authorities and perhaps also to the public, for the time being under the ePrivacy Directive but in the near future also in all personal data processing instances.[98] Consequently, the data protection framework treats the matter of security of processing seriously—the only flexibility afforded to data controllers pertains to the fact that security measures may be proportionate to the risks "*represented by the processing and the nature of the data to be protected*".

System security is also important in the smart grid context. System vulnerabilities may affect anything from individual data protection to state security. Accordingly, unlawful use of the system may constitute anything from data protection infringement to computer crime or state security matter. The above are highlighted only in order to demonstrate that the security of the processing is in any case a central concern of all parties involved in the provision of relevant services.

Given the omnipresent security concerns in the smart grid process, it is to be expected that data protection issues will be addressed within the general security policies adopted. The principle of proportionality is expected to hold a helpful role for system providers too: because the nature of personal data processed is not expected to be particularly threatening to individuals (unless of course such sensitive data are collected or may be processed as health data), the security measures adopted in order to address the, much higher, systemic risks (control of the state grid system, computer crime, state security) are expected to be enough to cover data protection concerns as well. At any event, it is likely that such data security accessories as impact assessments or security policies might be required from data controllers at various stages while implementing their smart grid systems.

---

[97] Art 17(2)–(4).
[98] Cf. Arts 31–32 of the EU General Data Protection Regulation.

2.7 The Non-binding EU Regulatory Framework for Personal Data Protection in Smart Grid and Smart Metering Systems

2.7.1 Opinions and Recommendations

Having overviewed, first, the rationale of EU regulation of smart meters and smart grids systems (Sect. 2.2) and, second, its general regulatory framework for these systems (Sect. 2.3), it is important, third, to analyse the *specific* EU regulatory framework for protecting personal data in smart grids and smart metering systems.

The policy makers at both the EU and its Member States level relatively early observed that privacy and data protection issues raised by the deployment of smart grid and smart metering systems must be appropriately addressed. From 2010 onwards, personal data protection became an equally important concern as cost-benefit analysis, technical issues, cyber-security or environmental protection, among others.

As of the time of writing, at the EU level, there is no binding legal instrument dealing with the protection of personal data in smart grid and smart metering systems. The EU opted for a "light" regulatory approach in which guidance is offered, in a first place, to the EU Member States and, secondly, to the data controllers and processors operating in the EU. The effectiveness of this approach might be questionable.

The International Working Group on Data Protection in Telecommunications ("the Berlin Group") and the Art 29 Working Party were among the first ones to express an interest in privacy and data protection in the context of smart grid and smart metering systems. As early as 2011, in parallel to the entry into force of the Third Energy Package (cf. *supra*, at 2.3.1.2), the latter has issued a seminal opinion in the field. It is one of the first instruments spelling out the legal issues in context of personal data protection in Europe that are raised by smart grid and smart metering systems.

International Working Group on Data Protection in Telecommunications, *Privacy by Design and Smart Metering: Minimize Personal Information to Maintain Privacy* **(2011)**[99]

1. Smart metering initiatives should feature privacy principles in the overall project governance framework and proactively embed privacy requirements into their design, in order to prevent privacy-invasive events from arising.

Utilities should conduct Privacy Impact Assessments (PIAs) or similar type assessments as part of the requirements and design stages of smart metering initiatives. Within this evaluation, two important considerations should be made. First, utilities should make a determination of what smart meter-based information is required to meet legitimate objectives

---

[99] Cf. http://www.datenschutz-berlin.de/attachments/842/675.43.18_WP_Privacy_and_Smart_Metering.pdf.

(and at what level of identifiability), rather than of what information is made available by smart metering. Mechanisms should then be put in place to allow consumers to maintain control over any available, non-necessary information. Secondly, only the personal information necessary for the determined purposes should leave the consumer's home via the smart meter. In order to ensure that consumers always retains control over their data it is essential that they are fully informed about the data which leave their homes. They should have the possibility to determine which data is sent and to intervene if necessary.

Some research has shown that utilities may not need detailed energy consumption information about individual consumers to perform load balancing functions. To achieve as little personal data flow as possible utilities may use techniques such as anonymisation, pseudonymisation, or data aggregation. Local gateways for individual buildings or small neighbourhoods, which allow the consumer to gain insight into their energy usage without the need for transmission of information about identifiable consumers to the utility, should be applied. Such gateways should generally not be externally- accessible and work with defined access protection profiles, while communication should be push- based (initiated by the gateway). Other measures, such as larger intervals between individual readings, can also prevent a detailed profile about the consumer's life-style from being generated. Of course, high technical standards for securely storing and accessing the data will be essential.

2. Smart meters should ideally protect privacy by default, with no action required on the part of the consumer.

3. Privacy should be an essential design feature of smart meter systems and practices.

... However, privacy cannot be solely reliant on legislative or administrative protections; it should also be designed into the technology itself. ...

4. Smart metering initiatives should avoid unnecessary trade-offs between privacy and other legitimate functionalities or organizational objectives.

... Consumers should not be forced into a choice between privacy and energy efficiency/conservation ...

5. Privacy and data security should be maintained end-to-end—full lifecycle protection.

Smart meter-based information—particularly personally identifiable information—should be strongly protected, whether at rest or in transit. This requires the development and implementation of data protections at the smart meter itself (ensuring, to the extent possible, that the device is tamperproof, and that it does not store more data than necessary), during transmission of the data (encryption, anonymisation, identification and protection of metadata), and during processing and use (minimized access to data, ensuring third parties meet equivalent protection standards, secure destruction at end-of-life, etc.).

6. Smart metering initiatives should be visible and transparent, and should utilize accountable business practices; consumers should be assured that the technology operates in accordance with stated objectives.

Utilities should be able to show that the methods used to incorporate privacy into their smart metering initiatives will meet the privacy requirements of the project. Ensuring such "requirements traceability" between the foundational privacy principles and each stage of a smart metering initiative will ensure that the utility is ready for a third part audit at any time.

Informing consumers of the use to which personal information collected from smart meters will be put, and the establishment of a clear and accessible complaints process, are key objectives in achieving visibility and transparency. Consumers should be given the simple technical option to define access control profiles and thus determine who receives what personal information.

7. Smart metering initiatives should be designed to respect consumer privacy—keep it user-centric.

Consumers should be provided with, and educated about, all necessary information, options and controls to allow them to manage their energy consumption and their privacy.

8. Regulatory frameworks should foster the introduction and use of privacy-friendly smart meter and smart grid applications.

The concepts laid out in the recommendations above should be incorporated into national and international regulatory frameworks where this is not already the case.

Art 29 Working Party, *Opinion 12/2011 on smart metering*, WP 183 (2011)[100]

Application of data protection law to the processing of data collected via smart meters

Where personal data are contained in the information generated and disseminated by a smart meter, the Working Party determines that Directive 95/46/EC applies to such processing.

From the general information available on this subject and from detailed discussions at national level with regard to the operation of smart meters, it has been established that the following data types can be assumed to be processed:

- Unique smart meter ID and/or unique property reference number (even in the absence of these identifiers, the meter might also be identified by its unique energy load graph);
- Metadata referring to the configuration of the smart meter;
- A description of the message being transmitted, for example whether it is a meter reading or a tampering alert;
- A date and time stamp;
- Message content.

Message content is likely to include the following types of information:

- Meter register read. This could be a single reading or a group of readings for a more complex tariff;
- Alerts. The meter may transmit a message informing that an event has triggered the meter's alarm;
- Network level information such as voltages, power outages and power quality;
- Load graphics with various levels of detail.

Data can be sent to the controller in real-time or be stored in the smart meter. In both cases however, under the Data Protection Directive, it is considered that the data have been collected by the controller.

---

[100] Cf. http://ec.europa.eu/justice/policies/privacy/docs/wpdocs/2011/wp183_en.pdf.

This list is far from exhaustive but the Working Party notes that the operation of smart meters—and by extension any further developments of smart grids and appliances—entail the processing of personal data as defined by Article 2 of Directive 95/46/EC and interpreted by the Working Party in its opinion 4/2007. Furthermore, the increased amount of personal data being processed, the possibility of remote management of connection and the likelihood of energy profiling based on the detailed meter readings make it imperative that proper consideration is given to individuals' fundamental rights to privacy.

The conclusion that personal data are processed has been reached for the following reasons:

1. the data enumerated above as being generated by smart meters is in most cases associated with unique identifiers such as a meter identification number. For domestic consumers of energy suppliers, this identifier is inextricably linked with the living individual who is responsible for the account. In other words, the device enables that individual to be singled out from other consumers;
2. further, the information collected in the context of a smart metering service relates to a consumer's energy profile in the context of their energy use and it is used to take decisions directly affecting that individual. Most obviously such a decision would be to determine the level of any charges for energy supply but it is not limited to billing purposes;
3. this view is further confirmed if one takes into account the widely promoted benefits of smart meter implementation such as the reduction of overall energy consumption in member states. Clearly, such an objective can only be achieved insofar as the energy consumption of individual consumers is also reduced and, according to energy suppliers and networks, achieving this objective is to a large extent dependent on the collection of large amounts of information about the behaviour of these consumers.

The definition of data controller as it applies to smart meters

It is established that the Directive 95/46/EC places obligations on the data controller with regard to their processing of personal data. Before setting out how those obligations apply in the context of this opinion, it is important for the Working Party to set out its view on which legal persons fall under the definition of data controller.

Smart meter implementation involves a number of organisations in the processing of personal data potentially including, but not limited to, energy suppliers, energy network operators, regulatory bodies, government bodies, third party service providers and communications providers. Given the number and complexity of relationships, it is likely that there will be difficulties in applying the relevant definitions but the analysis in this Opinion reflects the approach taken by the Working Party in its Opinion 1/2010 on the concepts of data controller and data processor. Therefore, the responsibilities stemming from data protection legislation should be clearly allocated in such a way that compliance with the data protection rules will be sufficiently ensured in practice.

Energy suppliers

In some Member States, the legal person with the most responsibility for processing personal data would be the supplier. They have the contract with the data subject which initiates the processing and by deciding which data they require to fulfil their functions and how they will collect, store and use the data, they can obviously be said to have determined the purposes for which, and the manner in which, the personal data are processed. This establishes them quite clearly as a data controller for the processing of personal data generated by an energy meter and the Working Party is of the view that, notwithstanding the added complexities brought about by smart meters, suppliers remain a data controller in this context.

Network Operators or DSOs

In other models, the DSO which owns the grid will be responsible for the installation and running of the Smart Meter system. The DSO will also be responsible for determining how the data are collected, stored and used. In this model the DSO will be a data controller. Where the energy suppliers have the right to access the data transmitted by the meters and are using the data for their own purposes (for example, to issue bills or to give advice to consumers) then they will also be a data controller for the personal data they are processing.

Other parties

There are many other parties who could potentially be processing personal data in the course of fulfilling their role in a programme to implement smart meters. Some of them may not even come into existence until the full effects of the shift towards greater amounts of personal data processing are apparent so it would not be prudent to attempt a definitive list at this stage. It is also relevant to remember the variations in supply models and concepts across member states. However, it is important to recognise that without all parties operating on a shared understanding of how the definition of data controller applies there is an increased risk that compliance and good practice will not be achieved. With this in mind the Working Party would remind all parties of the following important points:

1. In some implementation models a central communications function is established which has responsibility for managing the transmission of data between the meter and the supplier. It is possible that this function could exist as a data processor acting only on the instructions of the suppliers to and from whom it sends and receives data. However, if in any case the communications function is engaged in deciding whether personal data can be disclosed to a third party, or whether such data can be processed for new purposes, then the communications function could assume the role of data controller in respect of that personal data processing.
2. Energy Regulators are also important actors. They may have access to data for policy setting and research purposes. Insofar as those data are personal data then clearly the regulatory body will assume the role of a data controller.
3. Third party service providers (often referred to as Energy Service Companies or ES-COs) will have an increasingly prominent role in the use of data generated by smart meters. Where personal data are disclosed to the ESCO in order for them to provide a service either to the consumer or to another party, such as a supplier, then the ESCO will assume the role of a data controller.

Lawfulness of processing and legitimate grounds/purposes for processing

Once it has been established that a legal person is to be considered as a data controller, it is then important to set out the legal requirements placed on the data controller by the Data Protection Directive. In accordance with Article 6 of the Directive, personal data must be processed fairly and lawfully. For any personal data processing to be lawful it needs to satisfy one or more of the six grounds for legitimate processing set out in Article 7 of the Directive.

The Working Party notes that in many, if not all, member states the exact nature of the purposes for the processing of personal data stored on or transmitted by a smart meter has yet to be made absolutely clear or properly defined. In light of this, the Working Party would advise that such purposes need to be established before any claim that the grounds for processing are legitimate can be made. The Working Party also notes that each separate purpose has to be, in and of itself, legitimate and that one legitimate purpose cannot serve to further legitimise any other. Specifically, personal data cannot be reprocessed for another purpose which is incompatible with the purpose for which they were originally collected.

The Working Party's view is that there are five possible grounds for processing available to data controllers in this context.

Consent

It is clear that many of the purposes for which personal data may be used will relate to enhanced services offered to the data subject, such as time of use tariffs or energy advice. Where a data subject has agreed to accept such a service, it is likely that the service provider —either a supplier or a third party—will have the opportunity to gain the consent of the data subject for the processing of personal data.

The Working Party would remind data controllers that reliance on consent will require consideration of the fact that valid consent only exists when the data subject has made a fully-informed decision. It is not possible to use consent as a grounds for processing personal data unless the data subject has been given sufficient information about the personal data processing to make a genuine choice. In particular, where there are a number of different functionalities, then the consent should be granular enough to reflect these multiple purposes rather than one consent being used to legitimise possibly divergent and unrelated different purposes.

The Working Party would recommend that industry develops effective and practical means by which data subjects can express their consent. It is important to remember that consent has to be freely given and must therefore be capable of being revoked so the methods for gaining consent should build in the capability for the data subject to change his mind without going to excessive amounts of trouble. One possible solution could be to design the household control panel to include 'push button' consent. The availability of this type of functionality would depend on the sophistication of the design of the meter and control panel in order to ensure that the process of consent remains valid.

Contract

Processing may also be necessary for the performance of a contract to which the data subject is a party, or in order to take steps at the request of a data subject prior to entering into a contract. This legal basis could be used to legitimise the processing of personal data for the purpose of billing as without an accurately compiled bill, the contract to supply energy cannot be fulfilled. In respect of billing, it is important to remember the element of necessity in this condition. In other words, if the grounds for processing are for the performance of a contract that only requires the customer to be provided with and pay a quarterly bill, it is not necessary for the supplier to collect more frequent readings in order to fulfil that contract. Either the contract would need to include valid and legal provision for more frequent readings or the supplier would need to rely on another legal basis for those readings.

Performance of a task carried out in the public interest or in the exercise of official authority

In some member states, the network electricity operator is responsible for the performance of the physical network, but also for reducing the global electricity consumption. This electricity consumption concerns both the global electricity consumption, and the consumption during the peak hours. Those tasks are carried out in the public interest, and they legitimise the installation of the smart meters.

Legal obligation

In some member states, the network operator has the obligation of installing and collecting data through the smart meters for every new installation.

Legitimate interests

According to Article 7(f) of the Directive, the processing could be lawful if it is necessary for the purposes of the legitimate interests pursued by the data controller or by a third party or parties to whom the personal data are disclosed except where such interests are overridden by the interests or fundamental rights of the data subject.

The key point to be made here is that reliance on this legal basis depends on giving proper weight to the interests and rights of data subjects. It might seem inarguable that the legitimate interests of the data controller and society as a whole would be served by increased efficiency in energy supply and consumption and that this might be achieved via the personal data collected from smart meters. However, simply because this particular use of personal data seems legitimate (and, to many people, desirable) does not mean that it can be applied to legitimise every element of processing. In other words, the imperative to reduce energy consumption, although it might be a sensible public policy objective, does not override data subjects' rights and interests in every case.

Indeed, it is clear that including practical measures such as Privacy Enhancing Technologies and Privacy Impact Assessments to enhance the security and privacy of the data processed by smart meters will make it more likely that this condition for processing could be available to a data controller.

This is particularly important where processing for a data controller's legitimate interests is both inherently and disproportionately intrusive or where the effect of the processing is to cause unwarranted detriment to the data subject. Examples might include the creation of detailed profiles of data subjects that are, in fact, not needed to achieve the purpose, passing details to third parties without the knowledge or consent of the data subject, or the use of personal data to take decisions about remote disconnection without proper regard for an individual's data protection and other rights.

The Working Party would also remind industry that in some member states the possibility for the data subject to object to installation of the smart meter exists and that in such cases the data subject's preferences override any other interests.

Further compliance issues raised by smart metering

The Working Party recalls its opinion 168 in which it was stated that services and technologies which rely on the processing of personal data should be designed with privacy by default settings. In this respect, smart metering implementation should take place with privacy built in at the start, not just in terms of security measures, but also in terms of minimising the amount of personal data processed. Some member states have proceeded with implementation plans which require a Privacy Impact Assessment and the Working Party would recommend this approach.

The smart meters currently being tested in some member states collect several readings, depending on the type of contract to which the customer has subscribed. For example, if the customer has a simple contract in which he pays the same price for electricity throughout the day, the meter will collect a daily single reading. Alternatively, if the customer has a contract for which there are different prices depending on the time of day, the meter will be collecting ten different readings every day. At its most basic level, Privacy By Design would ensure that meter readings are only transmitted as frequently as necessary for the operation of the system or the provision of a service the consumer has agreed to receive. [...]

The technical specifications of the network should also ensure that any data collected should remain within the household network unless transmitting it elsewhere is necessary, or if the data subject consents to the transmission. Also, the system should be designed to ensure that even where personal data are transmitted, any data elements which are not necessary to fulfil the purpose of the transmission are filtered out or removed. The overall aim should be that the lowest possible data volumes are processed and transmitted.

The Working Party also recommends that systems are designed so as to allow access to personal data only to the extent necessary for the role being performed by the data controller. All parties who are accessing personal data should be verified as being appropriate and competent recipients of the personal data and they should only be capable of accessing personal data necessary for them to fulfil their role. They should not have access to personal data beyond this scope.

Retention of personal data

In the 'pre-smart' world, the energy industry has developed practices for the retention of personal data for a limited number of purposes, for example, billing. The smart metering environment presents new challenges. Given that substantially greater quantities of data will be processed, retention policies and practices will need to be established for new purposes and reviewed for existing purposes. In order to be certain that data is being retained only as long as necessary to achieve a specified and lawful purpose, then a clearer understanding of the purposes of processing must be established. This in turn will enable controllers to demonstrate that personal data is only being retained for as long as necessary. For example, one purpose mentioned quite frequently is that the data collected from a meter would allow for the provision of energy efficiency advice. In some cases this type of service might include offering year on year comparisons and it has been suggested that 13 months may be an appropriate time period for retaining personal data in order to satisfy this purpose. However, such a long retention period would only be acceptable where the data subject has agreed that they would take advantage of such a scheme. For the provision of other types of service a much shorter retention period should be required.

Furthermore, it is conceivable that consumers could hold much of this data on the meter or comparable gateway device (other than that required for billing purposes). This gives the opportunity to allow the data subject to make their own decisions regarding retention. If this were the case it would be advisable for consumers to receive a system of prompts or reminders to assist with this housekeeping.

Third Parties' processing of personal data

It is likely that there will be significant involvement of third parties/ESCOs delivering and supporting the smart metering implementation and the Working Party believes this will require careful consideration. The influence and involvement of third parties will vary from member state to member state but it is clear that at its most intrusive the implementation of smart metering could result in a trade in energy profiles in the interests of those parties wishing to market energy services.

Techniques that have been suggested to assist with compliance include establishing a central information and communications hub which acts as a conduit for all those involved who wish to access consumer data; a Code to which all parties must be signatories; and a Charter which would span the whole industry. The Working Party wishes to make it clear that the more intrusive the processing, the more rigorous the safeguards need to be. The Working Party would strongly urge relevant regulatory bodies to take a view on the acceptability of the more intrusive processing.

Underpinning all of these would be consumer consent, with the industry ensuring that the data subject is in a position to grant this in an informed way. The Working Party wishes to make it clear that it would be unacceptable for third parties to be processing detailed information about a data subject's energy usage without the knowledge and consent of that data subject.

Security

As part of the Privacy by Design process, security and privacy risk assessments will identify the potential risks to data security. Given the novel and vast prospect that is in store with the smart grid and its associated technologies, the task of anticipating security requirements is a challenging one.

Bearing this in mind, this Opinion recommends that in order to mitigate risk, the approach should be end-to-end, incorporating all parties and drawing on a broad range of expertise. Security should also be designed in at the early stage as part of the architecture of the network rather than added on later.

The Working Party wishes to make it clear that for data subjects to be confident that their personal data are processed securely and their fundamental right to privacy protected, appropriately robust security safeguards must be in place. These safeguards should apply to the whole process including the in-home elements of the network, the transmission of personal data across the network and the storage and processing of personal data by suppliers, networks and other data controllers.

The Working Party anticipates that smart meters will have a long life expectancy and therefore advises that safeguards will need to be updated and improved over time and must be regularly subject to review and testing.

Given the increased amounts of personal data being processed it is clear that the risk to data privacy also increases. Therefore, the Working Party recommends that technical and organisational safeguards should cover at least the following areas:

- The prevention of unauthorised disclosures of personal data;
- The maintenance of data integrity to ensure against unauthorised modification;
- The effective authentication of the identity of any recipient of personal data;
- The avoidance of important services being disrupted due to attacks on the security of personal data;
- The facility to conduct proper audits of personal data stored on or transmitted from a meter;
- Appropriate access controls and retention periods;
- The aggregation of data whenever individual level data is not required.

Individual rights including information provided to data subjects

The implementation of smart meters will give rise to complex and novel personal data processing operations. Most data subjects will be unaware of the nature of these operations and the potential impact this could have on their privacy. Certainly, if they are not aware of the personal data processing then it is impossible for them to make informed decisions about it. The obligation to inform data subjects about the processing of their personal data is one of the fundamental principles of the Data Protection Directive. Article 10 regulates the provision of this information and requires data controller to make the following information available to the data subject:

- The identity of the data controller and of his representative if any;
- The purposes of the processing;
- Any further relevant information which would render the processing fair. This includes the identity of the recipients of the personal data, the existence of the rights of access and rectification.

The data controller responsible for the installation and maintenance of the meter should make clear to data subjects what information is collected from the meter and what it is used for.

Insofar as third parties are involved in the processing of personal data for the purpose of providing services to data subjects, data subjects should be similarly informed. In some circumstances, it might be appropriate to allow for independent vetting or monitoring of third party access to and use of personal data to ensure that data subjects are not being misled.

Rights of the data subject

Data controllers must respect the rights of data subjects to access and, where appropriate, to correct or delete information held about them. Clearly, the fact that an integral part of the smart metering project is the implementation of a 'home network' (where the consumer can obtain instant information from the smart meter about their usage patterns and tariffs), means that there is an opportunity to ensure that data subjects are able to exercise their rights easily using tools that enable direct access to data. [...]

Processing of data for crime prevention and investigation

The Data Protection Directive regulates against the processing of personal data where the processing is excessive with regard to the purpose. It is clear that the detailed picture obtained by smart meters that inform suppliers about patterns of energy use might allow for the identification of suspicious and, in some cases, illegal activities. The Working Party would remind the industry that the fact that such a possibility exists does not automatically legitimise wide scale processing of data for this purpose. It is particularly important to note that insofar as personal data relates to the alleged committing of an offence, that this personal data would be categorised as being sensitive and, as a result, the data controller cannot process such data unless Article 8 (5) of the Directive applies.

Conclusion

The arrival of smart metering, which paves the way for the smart grid, brings with it an entirely new and complex model of inter-relationships which poses challenges for the application of data protection law. [...]

This opinion explains the applicability of data protection law: it has demonstrated that personal data is being processed by the meters, so data protection laws apply. [...]

Whatever the processing, whether it is similar to that which existed in the pre-smart environment, or unprecedented, the data controller must be clearly identified, and be clear about obligations arising from data protection legislation including Privacy by Design, security and the rights of the data subject. Data subjects must be properly informed about how their data is being processed, and be aware of the fundamental differences in the way that their data is being processed so that when they give their consent it is valid.

The Art 29 Working Party opinion was followed by a normative instrument, namely the European Commission's 2012 Recommendation.[101] Personal data protection constitutes one of the three main blocks thereof (cf. *supra*, at 2.3.2). To that end, the 2012 Recommendation suggests five main "tools" for achieving the adequate level of protection:

- data protection impact assessment (DPIA),
- data protection by default,
- data protection by design,
- privacy enhancing technologies (PETs),
- best available techniques (BATs).

---

[101] Cf. *supra*, note 41.

European Commission, *Recommendation on the roll-out of smart grid and smart metering systems* **(2012)**[102]

Data protection impact assessments

4. The data protection impact assessment should describe the envisaged processing operations, an assessment of the risks to the rights and freedoms of data subjects, the measures envisaged to address the risks, safeguards, security measures, and mechanisms to ensure the protection of personal data and to demonstrate compliance with Directive 95/46/EC, taking into account the rights and legitimate interests of data subjects and persons concerned.

5. In order to guarantee protection of personal data throughout the Union, Member States should adopt and apply the data protection impact assessment template to be developed by the Commission and submitted to the Working Party on the protection of individuals with regard to the processing of personal data for its opinion within 12 months of publication of this Recommendation in the Official Journal of the European Union.

6. When implementing this template, Member States should take into account the advice of the Working Party on the protection of individuals with regard to the processing of personal data.

7. Member States should ensure that network operators and operators of smart metering systems, in line with their other obligations under Directive 95/46/EC, take the appropriate technical and organizational measures to ensure protection of personal data.

8. Member States should ensure that the entity processing personal data consults the Data Protection Supervisory Authority referred to in Article 28 of Directive 95/46/EC on the data protection impact assessment, prior to processing. This should allow the authority to assess the compliance of the processing and, in particular, the risks for the protection of personal data of the data subject and the related safeguards.

9. Member States should make sure that once the template for data protection impact assessments, as provided for in point 5, has been adopted, network operators implement the points 7 and 8 in accordance with it.

Data protection by design and data protection by default settings

10. Member States should strongly encourage network operators to incorporate data protection by design and data protection by default settings in deployment of smart grids and smart metering.

11. Data protection by design and data protection by default settings should be incorporated in the methodologies of parties involved in development of smart grids when personal data are processed.

12. Data protection by design should be implemented at legislative level (through legislation that has to be compliant with data protection laws) at technical level (by setting appropriate requirements in smart grid standards to ensure that infrastructure is fully consistent with the data protection laws) and organizational level (relating to processing).

13. Data protection by default should be implemented so that the most data protection friendly option is provided to the customer as a default configuration.

---

[102] Cf. *supra*, note 41.

14. Member States should encourage European standardization organizations to give preference to smart grid reference architectures based on data protection by design and on data protection by default.

15. For the purposes of optimizing transparency and the individual's trust, Member States should encourage use of appropriate privacy certification mechanisms and data protection seals and marks, provided by independent parties.

16. Article 8 of the Charter of Fundamental Rights of the European Union and Article 8(2) of the European Convention on Human Rights require justifying any interference with the right to the protection of personal data. The legitimacy of interference must be assessed on a case-by-case basis in the light of the cumulative criteria of legality, necessity, legitimacy, and proportionality. Any processing of personal data which interferes with the fundamental right to the protection of personal data within the smart grid and smart metering system therefore has to be necessary and proportional for it to be considered fully in compliance with the Charter.

17. In order to mitigate the risks on personal data and security, Members States, in collaboration with industry, the Commission and other stakeholders, should support the determination of best available techniques for each common minimum functional requirement listed in point 42 of the Recommendation.

Data protection measures

18. When deciding the range of information allowed for processing within smart grids, Member States should take all necessary measures to impose, as much as possible, use of data rendered anonymous in such a way that the individual is no longer identifiable. In cases where personal data are to be collected, processed, and stored, Member States should ensure that the data are appropriate and relevant. Data collection should be limited to the minimum necessary for the purposes for which data are processed and data should be kept in a form which permits identification of data subjects for no longer than is necessary for the purposes for which the personal data are processed.

19. Processing of personal data by or within a smart metering system should be legitimate in accordance with one or more of the grounds listed in Article 7 of Directive 95/46/EC. The opinion of the Working Party on the protection of individuals with regard to the processing of personal data on smart metering should be taken into account.

20. The processing of personal data by third parties offering value-added energy services should also be lawful and based on one or more of the six grounds for legitimate processing listed in Article 7 of Directive 95/46/EC. Where consent is chosen as the ground for processing, the consent of the data subject should be freely given, specific, informed, and explicit and be given separately for each value-added service. The data subject should have the right to withdraw his or her consent at any time. The withdrawal of consent should not affect the lawfulness of the processing based on the consent before the withdrawal.

21. Member States should clearly determine the roles and responsibilities of data controllers and data processors. They should be compatible with their respective obligations set out in Directive 95/46/EC.

22. Member States should perform an analysis prior to launching processing operations, in order to determine to which extent suppliers and network operators need to store personal data for the purposes of maintaining and operating the smart grid and for billing. This analysis should allow Member States to determine, *inter alia*, if the periods for the storage of personal data currently set in national law are no longer than necessary for the purposes of operating smart grids. This must include mechanisms to ensure that the time limits set for the erasure of personal data and for a periodic review of the need to store personal data are observed.

23. For the purpose of this analysis, each Member State should particularly take into account the following principles: the principle of data minimization, the principle of transparency—by ensuring that the end consumer is informed in a user-friendly and intelligible form using clear and plain language, of the purposes, timing, circumstances, collection, storage, and all other processing of personal data, and the principle of empowerment of the individual—by ensuring that the measures taken safeguard the individual's rights.

Data security

24. Member States should ensure that personal data security is designed in at an early stage as part of the architecture of the network, within a data protection by design process. This should encompass measures to protect personal data against accidental or unlawful destruction or accidental loss and to prevent any unlawful forms of processing, in particular any unauthorized disclosure, dissemination, access to or alteration of personal data.

25. The use of encrypted channels is recommended as it is one of the most effective technical means against misuse.

26. Member States should take into account that all present and future components of smart grids ensure compliance with all the "security-relevant" standards developed by European standardization organizations, including the smart grid information security essential requirements in the Commission's standardisation mandate M/490. The international security standards should also be taken into account, in particular the ISO/IEC 27000 series ("ISMS family of standards").

27. Member States should ensure that network operators identify security risks and the appropriate security measures to guarantee the adequate level of security and resilience of the smart metering systems. In this regard, network operators, in cooperation with national competent authorities and civil society organizations, should apply existing standards, guidelines, and schemes and where not available develop a new one. Relevant guidelines published by the European Network Information and Security Agency (ENISA) should also be taken into account.

28. Member States should ensure that in accordance with Article 4 of Directive 2002/58/EC, in the event of a personal data breach, the controller notifies without undue delay (preferably not later than 24 h after the breach has been established) the supervisory authority and the data subject, if the breach is likely to have an adverse effect on protection of his or her personal data.

Information and transparency on smart metering

29. Without prejudice to the obligations of data controllers, in accordance with Directive 95/46/EC, Member States should require that network operators develop and publish an accurate and clear information policy for each of their applications. The policy should include at least the items mentioned in Articles 10 and 11 of Directive 95/46/EC.

Where personal data relating to a data subject are collected, the controller should also provide the data subject with at least the following information:

(a) the identity and the contact details of the controller and of the controller's representative and of the data protection officer, if any;
(b) the purposes of the processing for which the personal data are intended, including the terms and general conditions and the legitimate interests pursued by the controller if the processing is based on Article 7 of Directive 95/46/EC;
(c) the period for which the personal data will be stored;

(d) the right to ask the controller for access to and rectification or erasure of the personal data concerning the data subject or to object to the processing of such personal data;
(e) the right to lodge a complaint with the supervisory authority referred to in Article 28 of Directive 95/46/EC and the contact details of the supervisory authority;
(f) the recipients or categories of recipients of the personal data;
(g) any further information necessary to guarantee fair processing in respect of the data subject, having regard to the specific circumstances in which the personal data are collected.

The European Data Protection Supervisor (EDPS), a data protection authority for EU institutions and bodies, has immediately reacted to the adoption of the 2012 Recommendation, to the extent it deals with personal data protection.

Opinion of the European Data Protection Supervisor on the Commission Recommendation on preparations for the roll-out of smart metering systems (2012)[103]

9. The EDPS Opinion has three main objectives and messages:

- First, the Opinion evaluates the Recommendation: it welcomes the Recommendation as a first step, highlights its achievements, but also criticises its shortcomings, including its insufficient specificity.
- Second, while the EDPS regrets that the Recommendation has not provided more specific and more practical guidance on data protection, he considers that some guidance can still be given in the data protection impact assessment Template, which is currently under preparation. Therefore, the Opinion provides a number of targeted recommendations on the Template.
- Third, the Opinion calls on the Commission to assess whether, beyond the adoption of the Recommendation and the Template, further legislative action is necessary at the EU level and provides a number of targeted recommendations for possible legislative action.

[...]

22. Considering the risks to data protection, one of the key pre-conditions for the rollout of smart metering systems is to ensure a high level of protection of personal data.

[...]

24. The EDPS particularly appreciates the efforts of the Commission to make use of newly proposed concepts such as data protection by design and practical tools such as data protection impact assessments and security breach notifications. The EDPS also welcomes the references in the Recommendation to data minimization, data protection by default, privacy-enhancing technologies ('PETS's), transparency, and consumer empowerment.

25. The EDPS, in particular, supports the Commission's plan to prepare a template for data protection impact assessment and submit it to the WP29 for advice. This approach may help bring consistency and encourage the implementation of the data protection by design principle in the Member States.

---

[103] Cf. https://secure.edps.europa.eu/EDPSWEB/webdav/shared/Documents/Consultation/Opinions/2012/12-06-08_Smart_metering_EN.pdf.

[RECOMMENDATIONS:]

Recommendation and Template should not be read to reduce, in any way, the data protection safeguards established in applicable data protection law.

Data protection concerns should be adequately considered as part of the cost benefit analysis for the roll-out of smart metering systems.

Need for more specific and pragmatic guidance.

Need for close cooperation between energy regulators and data protection authorities

[...]

46. The EDPS recommends that a freely given, specific, informed and explicit consent must be required for all processing that goes beyond processing required for (i) the provision of energy, (ii) the billing thereof, (iii) detection of fraud consisting of unpaid use of the energy provided, and (iv) preparation of aggregated data necessary for energy-efficient maintenance of the grid (forecasting and settlement).

50. Finally, the EDPS would welcome further safeguards with regard to access to smart metering data by law enforcement, tax authorities, other government agencies, insurance companies, employers, and other third parties. The EDPS would recommend, among others, restricting law enforcement access to cases when a judicial warrant, or adequate legal instrument, has been obtained, similar to a search warrant before police may enter and search an individual's home.

[...]

Who should prepare a data protection impact assessment? The Template should also provide guidance to industry participants on who should prepare a data protection impact assessment. It should be ensured that all controllers responsible for the processing of personal data (for example, network operators as well as energy suppliers, and operators of the smart metering system but also providers of value added services) each carry out an impact assessment relating to their data processing activities.

- Impact assessments can be prepared individually by each party (e.g. individually by each energy supplier and each network operator). However, considering the complexity of the data flows and the multiplicity of controllers and processors, coordination of activities and exchange of best practice may be particularly important. When appropriate, it may even be useful, and may help development of best practice and avoid duplication of efforts if several parties prepare the impact assessment jointly. That said, clear allocation of liabilities is equally crucial, and joint efforts should not lead to lack of ultimate responsibility towards the data subjects.
- As regards devices (such as smart meters and in-home displays), to ensure a data protection by design approach, and in particular, that data protection will be taken into account in the design of the functionalities of the devices, the EDPS recommends that a data protection impact assessment also be carried out for each device.

2.7.2 Data Protection Impact Assessment Template for Smart Grid and Smart Metering Systems

Following the development of a privacy impact assessment (PIA) template for radio-frequency identification (RFID) applications,[104] in accordance with § 5 of the 2012 Recommendation (cf. *supra*, at 2.7.1), the Expert Group 2 (EG2) of the Smart Grids Task Force, empowered by a renewed mandate (cf. *supra*, at 2.4.2), developed between February 2012 and January 2013 the data protection impact assessment (DPIA) template for smart grid and smart metering systems. It was subsequently submitted to the Art 29 Working Party (WP29) for an opinion, who on 22 April 2013 judged it negatively. Although the WP29 found the "approach outlined in the proposed document is sound", it has identified three critical concerns, i.e. the lack of clarity on the nature and objectives of the DPIA, certain methodological flaws as well as the lack of sector-specific content. It further suggested the integration of best available techniques (BATs) into the DPIA process.

Art 29 Working Party, *Opinion 04/2013 on the Data Protection Impact Assessment Template for Smart Grid and Smart Metering Systems ('DPIA Template') prepared by Expert Group 2 of the Commission's Smart Grid Task Force* (2013)

2.1. *Lack of clarity on the nature and objectives of the DPIA*

[...] the objective of a DPIA should [...] be to assess the impacts of the risks on the data subjects.

The WP29, however, regrets that the submitted DPIA Template does not directly address the actual impacts on the data subjects, such as, for example, financial loss resulting from inaccurate billing, price discrimination or criminal acts facilitated by unauthorised profiling. Even if the data protection and privacy targets listed in Annex I can be very useful to facilitate compliance, they are not sufficient in the context of a risk driven approach. Assessment of the potential impacts on data subjects is an indispensable element of such an approach.

Therefore, the WP29 considers that the DPIA Template in its current form cannot achieve its objective mandated by the Commission Recommendation.

2.2. *Methodological flaws in the DPIA Template*

In addition to, and sometimes linked to, the key shortcoming identified above, the WP29 believes that the DPIA Template suffers from a number of methodological flaws that jeopardise its application.

Firstly, the proposed DPIA Template often confuses risks and threats.

Secondly, there is no matching between the risks to be mitigated and the list of possible controls in Annex II. Even if each risk scenario is specific and should be assessed in its peculiarity, it is often possible to identify certain categories of controls as being effective in

---

[104] Opinion 9/2011 on the revised Industry Proposal for a Privacy and Data Protection Impact Assessment Framework for RFID Applications, http://ec.europa.eu/justice/policies/privacy/docs/wpdocs/2011/wp180_en.pdf. Cf. also [15].

mitigating certain risk categories. […] Suggested mitigating measures, while not replacing the need for a risk driven process, can provide a reference for an effective and coherent approach. […]

In addition, the proposed DPIA Template also does not give enough detail and specific guidance on the concept of vulnerability, on how to calculate and prioritise risks, choose the appropriate controls and assess the residual risks that remain after the controls have been put in place. […] It is also not clear how to complete the proposed forms.

Finally, the DPIA Template does not provide sufficient advice on how to determine data protection roles and responsibilities of the different stakeholders. […] it seems critical to provide the industry with guidelines allowing the identification of data controllers and data processors. […]

2.3. The DPIA Template lacks sector-specific content: industry-specific risks and relevant controls to address those risks should be identified and matched

The DPIA Template lacks sector-specific content. Both the risks and the controls listed in the template are of generic nature and only occasionally contain industry-specific guidance —best practice that could be genuinely useful. In a nutshell: the risks and controls do not reflect industry experience on what the key concerns and best practices are. The WP29 understands that the EG2 is currently working on a collection of 'best available techniques' ('BATs'). […] the Commission should consider integrating the BATs into it and submit the integrated document to the WP29 for an opinion. […]

In addition, the notion of a DPIA template is different from the notion of a DPIA framework. A framework should identify objectives, outline a methodology and define the scope of the assessment in terms of the boundaries of the system/process under analysis. A template should go further and provide an operational instrument to manage the risks of the specific system/process and its use cases, suggest possible controls and best available techniques to mitigate those risks and provide specific guidance. This is particularly needed in cases where no specific expertise is at disposal (SMEs, for example, or as in the smart grid case, in an industry that has previously faced relatively few privacy and data protection issues).

The DPIA Template should aim at developing more sector-specific and easier to use guidance. In particular, it is necessary to better define potential impacts on the data subjects in the smart-grid context and to give more precise guidelines regarding the type of controls that can be implemented.

The Commission could have provided EG2 a generic privacy and data protection risk assessment methodology. EG2 could have, in turn, applied such a methodology, and based on such methodology, could have made the DPIA Template more sector-specific. This approach would have allowed EG2 to focus on relevant issues such as smart grid specific risks and controls while relying on the reference framework for fundamental methodological aspects. […]

3. Conclusion and recommendations

The WP29 […] is of the opinion that the DPIA Template in its current form is not sufficiently mature and well-developed. […]

Given the identified shortcomings of the DPIA Template, the WP29 further recommends that the Commission should consider integrating the BATs into the DPIA Template and submit the integrated document to the WP29 for an opinion.

Further, and more broadly, the WP29 recommends the Commission to consider taking stock of past and on-going work in the field of DPIAs and to consider the opportunity of defining a generic DPIA methodology from which field specific efforts could benefit.

Finally, with regard to the need for a mandatory impact assessment, the WP29 refers to the experience gained with the RFID [PIA] and emphasises that available statistics in Member States shows that the take-up of impact assessments for RFID has been extremely low. Whereas these statistics may have several underlying reasons, one of the key contributing factors definitely appears to be the current lack of a mandatory requirement to carry out such an impact assessment.

This development has lead the EG2 to a significant improvement of the template, which on 20 August 2013 was submitted to the WP29 and on 4 December 2013 received a positive judgement. The WP29's second opinion has recognized "the work carried out by the EG2 group and realises that the second version of the template constitutes considerable improvement with respect to the previous version". The opinion is primarily devoted to a number of shortcomings in the risk management methodology, rather technical in their nature, but it addresses also the question of the effectiveness of a non-binding framework, the involvement of national data protection authorities in the DPIA process and the need for a test phase to assess the usefulness and efficiency of the template.

Art 29 Working Party, *Opinion 07/2013 on the Data Protection Impact Assessment Template for Smart Grid and Smart Metering Systems ('DPIA Template') prepared by Expert Group 2 of the Commission's Smart Grid Task Force* (2013)

2.1.1. *On the discretional nature of performing a smart grid DPIA*

The existence of a Commission Recommendation, while on the one hand not imposing a legally binding obligation, on the other hand sets forth that certain measures are strongly recommended. [...] The WP29 wants to reaffirm that the need for (a DPIA) [...] is largely justified by the complexity of smart grids technical and management infrastructure, by its potential scale of application and evolution, and by the specific risks for the individual's fundamental rights and freedoms, including, among others, life (e.g. switch off of energy supply where certain powered machines support vital functions).

Furthermore, the WP29 has welcomed the fact that the Commission has proposed a General Data Protection Regulation that would make data protection impact assessments mandatory under certain conditions. It should be clear for the stakeholders of the Smart Grid DPIA template, i.e. data controllers and processors, that the use of the template should be seen as a means to comply with a legal obligation in the future.

2.1.2. *The DPIA and the Data Protection authorities*

Point 8 of the Commission Recommendation provides that Member States should ensure that the entity processing personal data consult their DPAs on the data protection impact assessment, prior to processing. The WP29 notices that the template is not fully reflecting this approach in many parts. [...]

2.5. *Need for testing/validation of the DPIA template*

The WP29 suggests that an adequate certain testing/validation of the DPIA Template be carried out, on the field on the basis of the existing version, and taking as much as possible account of the above comments. The WP29 suggests that following these test, the template and its methodology should be reviewed and enhanced in the light of those experiences and taking into account the aforementioned comments.

After some minor editorial changes, the DPIA template was concluded on 10 March 2014 and was made public in October 2014.[105] The DPIA template consists of three sections. First, it overviews the rationale, scope, benefits, success factors of the DPIA process as well as it discusses the stakeholders to be involved in such a process. These include transmission system operators (TSOs), distribution system operators (DSOs), energy generators, energy market suppliers, metering operators, energy services organizations as well as—to a certain extent—consumers (i.e. data subjects).

Second, the template offers a detailed guidance on preforming the DPIA, fore-seeing the following steps:[106]

Exhibit 3: The DPIA process for smart grids and smart metering systems

Step 1—Pre-assessment and criteria determining the need to conduct a DPIA

    Criterion 1: Processing of personal data
    Criterion 2: Classification of data controllers and data processors
    Criterion 3: Impacts on rights and freedoms
    Criterion 4: Timing and motivation to perform a DPIA
    Criterion 5: The nature of the system/application
    Criterion 6: Legal basis and public concerns
    [other]

Step 2—Initiation

    The DPIA team
    Resources needed

Step 3—Identification, characterisation and description of Smart Grid systems/ applications processing personal data

    The use case
    System information
    Description of primary and supporting assets of the system

Step 4—Identification of relevant risks

    Threats identification for each feared event

Step 5—Data protection risk assessment

    Impact of feared events
    Likelihood of threats
    Final risk level/value priority

Step 6—Identification and Recommendation of controls and residual risks

    Assessment of implemented and planned controls
    Risk treatment

(continued)

---

[105] The two previous drafts of the DPIA template were never officially made public. The final version is published online, cf. http://ec.europa.eu/energy/gas_electricity/smartgrids/doc/2014_dpia_smart_grids_forces.pdf.

[106] ibid, pp. 14–35.

Residual risks and risks acceptance
Resolution

Step 7—Documentation and drafting of the DPIA Report

Step 8—Reviewing and maintenance

Third, the template offers a form that could be filled in while preparing the final report from the DPIA process. Furthermore, it offers a non-exhaustive list of data protection threats. Appended to the template are a glossary, a list of so-called privacy and data protection targets, and a list of possible controls.

The template suggests a particular risk management methodology, built on a relevant handbook by *La Commission Nationale de l'Informatique et des Libertés* (CNIL), the French data protection authority.[107] However, it does not preclude the application of other methodologies.

The publication of the DPIA template was complemented by the European Commission's recommendation, specifically addressing the use of the DPIA template and its evaluation ("the 2014 Recommendation").[108] It invites the EU Member States to encourage data controllers to apply the DPIA template (§ 3), to stimulate and support its dissemination and use (§ 4), to complement its application with best available techniques (BATs) (§ 5) and to consult a national DPA on DPIA, prior to the commencement of personal data processing (§ 7). It next introduces a test phase in which the efficiency and efficacy of the current DPIA template will be evaluated (§§ 9–13). It introduces a public inventory of DPIAs actually conducted (§ 14). The Recommendation concludes by a revision clause (§§ 15–17).

Commission Recommendation of 10 October 2014 on the Data Protection Impact Assessment Template for Smart Grid and Smart Metering Systems (2014/724/EU)[109]

II. DEFINITIONS

2. Member States are invited to take note of the following definitions:

[...] (c) 'data protection impact assessment' means a systematic process for evaluating the potential impact of risks where processing operations are likely to present specific risks to the rights and freedoms of data subjects by virtue of their nature, their scope or their purposes to be to carried out by the controller or processor or the processor acting on the controller's behalf [...]

III. IMPLEMENTATION

3. In order to guarantee protection of personal data throughout the Union, Member States should encourage data controllers to apply the DPIA Template for Smart Grid and Smart Metering Systems, and in doing so, encourage them to take into account the advice of the Working Party

---

[107] Cf. http://www.cnil.fr/fileadmin/documents/en/CNIL-ManagingPrivacyRisks-Methodology.pdf.

[108] European Commission, Recommendation of 10 October 2014 on the Data Protection Impact Assessment Template for Smart Grid and Smart Metering Systems, 2014/724/EU, OJ L 300, 18.10.2014, pp. 63–68.

[109] Cf. *supra*, note 111.

on the protection of individuals with regard to the processing of personal data, in particular its Opinion 07/2013. The opinions of the Working Party are available on the Smart Grid Task Force's webpage (http://ec.europa.eu/energy/gas_electricity/smartgrids/smartgrids_en.htm).

4. Member States should cooperate with industry, civil society stakeholders and national data protection authorities to stimulate and support the dissemination and use of the DPIA Template at an early stage in the deployment of smart grids and the roll-out of smart metering systems.

5. Member States should encourage data controllers to consider as a complementary element to the Data Protection Impact Assessment, the Best Available Techniques to be determined by Member States in collaboration with the industry, Commission and other stakeholders for each of the common minimum functional requirements for electricity smart metering as listed in point 42 of Recommendation 2012/148/EU […].

7. Member States should ensure that the data controllers consult their respective national data protection authorities on the data protection impact assessment, prior to processing.

8. Member States should ensure that data controllers, following the conduct of a data protection impact assessment, and in line with their other obligations under Directive 95/46/EC, take the appropriate technical and organisational measures to ensure protection of personal data, and review the assessment and continued appropriateness of the identified measures throughout the lifecycle of the application or system.

IV. TEST PHASE

9. Member States should support the organisation of a test phase with deployment of real cases, including by seeking and encouraging testers from the smart grid and smart metering industries to engage in this test phase.

10. Member States should ensure, during this test phase, that all relevant applications or systems apply the Template, the advice of the Working Party on the protection of individuals with regard to the processing of personal data, as well as the provisions contained in Section III of this Recommendation, in order to have the best impact on data protection and to provide as much input as possible for the Template's subsequent review.

11. Member States should encourage and support national authorities competent for data protection to offer their support and guidance to data controllers throughout the test phase.

12. The Commission intends to directly contribute to the implementation and monitoring of the test phase by facilitating dialogue and cooperation amongst stakeholders, in particular by providing the stakeholder platform for the organisation of stakeholder meetings involving the testers, industry and civil society representatives, national data protection authorities and energy regulators.

13. Member States should encourage the testers to communicate and share the results of the test phase with the national authorities competent for data protection and with the other relevant stakeholders in the framework of the stakeholder platform based on three categories of evaluation criteria:

(a) efficiency of the Template in assessing the impact of individual smart grid applications on data protection;
(b) usefulness of the Template in guiding the data controller in the conduct of the impact assessment according to the concrete circumstances of the application or system; and
(c) user-friendliness of the Template from the data controller's perspective.

The reporting on these evaluation criteria should focus on providing information relevant to the application of the Commission Recommendation and of the Template across all relevant applications or systems.

14. The Commission intends to ensure the compilation of an inventory of data protection impact assessments conducted during the test phase. The inventory of data protection impact assessments will be made available on the Smart Grid Task Force's webpage throughout the test phase and will be regularly updated in order to foster continuous and prompt improvement in the Template's application.

V. REVIEW

15. Within two years of publication of this Recommendation in the Official Journal of the European Union, Member States should provide the Commission with an assessment report highlighting the relevant conclusions stemming from the test phase.

16. Two years after the publication of this Recommendation in the Official Journal of the European Union, the Commission intends to assess the need for revision of the DPIA Template based on the test phase reports provided by Member States and in light of the abovementioned evaluation criteria. The Commission will consider organising a dedicated stakeholder event to exchange views on this assessment prior to undertaking a revision.

17. This revision should contribute to ensure that the DPIA Template provides improved data protection to individuals in the context of the deployment of smart grids and adequately reflects the provisions of the revised Directive 95/46/EC and the Working Party's Opinion 07/2013.

2.8   Tools for the Protection of Privacy and Personal Data

Since 1990s, a number of so-called "tools" for the protection of privacy and personal data developed and matured. These constitute non-legal means (methodologies) supplementing the legal ones in achieving the goal of adequate protection of privacy and personal data.

2.8.1   Privacy by Default

Cavoukian, *Operationalizing Privacy by Design: A Guide to Implementing Strong Privacy Practices* (2012) [16]

The single most effective yet most challenging method of preserving privacy is to ensure that the default settings—the settings that apply when the user is not required to take any action—are as privacy-protective as possible. [...]

The starting point for designing information technologies and systems must always be maximally privacy-enhancing, beginning with NO collection of personally identifying information, unless and until a specific and compelling purpose is defined. [...]

Where personal data must be collected for clearly specified purposes, the next step in operationalizing this principle is to limit the uses and retention of that information, as much as possible. There are many ways in which this may be accomplished. One method is to carry out operations with privacy implications (i.e. those that use personal information) client-side—that is, entirely under the control of users and their devices. Obviously, the more tamper-proof, secure, and user-controlled the device or software, the more trusted it will be to carry out its functions reliably. Dividing data, functions, and roles among different entities is a proven method of ensuring privacy. For example, this strategy is the basis for using proxy servers to obscure IP addresses and to defeat online tracking and profiling. In practice, a combination of organizational and technical measures will be necessary to achieve this goal of default privacy. […]

The default principle is illustrated in the following examples: […]

6. Distributed Information Privacy Architecture: Separating Domains in Service-Oriented Architecture in the Smart Grid. Ontario's Hydro One utility used the concept of "Domains" to classify the possible implications for privacy in the Smart Grid and to impose certain architectural decisions to meet privacy requirements. The three domains identified were: Customer Domain, Service Domain, and Grid Domain. As a result of the analysis, the default designed into the energy utility's Advanced Distribution System (ADS) separated the various data domains and conducted data aggregation on consumer data in a dynamic manner.

2.8.2 Privacy by Design (PbD)

Cavoukian, *Privacy by Design* (2013) [17]

Privacy by design (PbD) was developed by the Information and Privacy Commissioner of Ontario, Canada, Dr. Ann Cavoukian, back in the '90s. Privacy by Design advances the view that the future of privacy cannot be assured solely by compliance with legislation and regulatory frameworks; rather, privacy assurance must become an organization's default mode of operation. […]

The 7 Foundational Principles of Privacy by Design have proven to be a valuable resource for individuals and organizations around the world. […] the 7 Foundational Principles of Privacy by Design have been translated into 31 official languages. The objectives of Privacy by Design—ensuring privacy protection and gaining personal control over one's own information and, for organizations, gaining a sustainable competitive advantage—may be accomplished by practicing the 7 Foundational Principles:

1. Proactive not Reactive; Preventative not Remedial

The Privacy by Design approach is characterized by proactive rather than reactive measures. It anticipates and prevents privacy invasive events before they happen. Privacy by Design does not wait for privacy risks to materialize, nor does it offer remedies for resolving privacy infractions once they have occurred—it aims to prevent them from occurring. In short, Privacy by Design comes before-the-fact, not after.

2. Privacy as the Default Setting

We can all be certain of one thing—the default rules! Privacy by Design seeks to deliver the maximum degree of privacy by ensuring that personal data are automatically protected in any given IT system or business practice. If an individual does nothing, their privacy still remains intact. No action is required on the part of the individual to protect their privacy—it is built into the system, by default.

3. Privacy Embedded into Design

Privacy by Design is embedded into the design and architecture of IT systems and business practices. It is not bolted on as an add-on, after the fact. The result is that privacy becomes an essential component of the core functionality being delivered. Privacy is integral to the system, without diminishing functionality.

4. Full Functionality—Positive-Sum, not Zero-Sum

Privacy by Design seeks to accommodate all legitimate interests and objectives in a positive-sum win-win manner, not through a dated, zero-sum approach, where unnecessary trade-offs are made. Privacy by Design avoids the pretense of false dichotomies, such as privacy versus security—demonstrating that it is possible to have both.

5. End-to-End Security—Full Lifecycle Protection

Privacy by Design, having been embedded into the system prior to the first element of information being collected, extends securely throughout the entire lifecycle of the data involved—strong security measures are essential to privacy, from start to finish. This ensures that all data are securely retained, and then securely destroyed at the end of the process, in a timely fashion. Thus, Privacy by Design ensures cradle to grave, secure lifecycle management of information, end-to-end.

6. Visibility and Transparency—Keep it Open

Privacy by Design seeks to assure all stakeholders that whatever the business practice or technology involved, it is in fact, operating according to the stated promises and objectives, subject to independent verification. Its component parts and operations remain visible and transparent, to users and providers alike. Remember, trust but verify.

7. Respect for User Privacy—Keep it User-Centric

Above all, Privacy by Design requires architects and operators to protect the interests of the individual by offering such measures as strong privacy defaults, appropriate notice, and empowering user-friendly options. Keep it user-centric.

[...] Privacy by Design provides a method for proactively embedding privacy into information technology, business practices, and networked infrastructures.

[...] Over the past several years, Commissioner Cavoukian has produced more than 60 *Privacy by Design* papers, written with many well-known subject matter experts including business executives, risk managers, legal experts, designers, analysts, software engineers, computer scientists, applications developers in telecommunications, health care, transportation, energy, retail, marketing, and law enforcement. While some of those papers are "foundational" works, much of the Privacy by Design research is directly related to one of nine key application areas:

1. CCTV/Surveillance Cameras in Mass Transit Systems;
2. Biometrics Used in Casinos and Gaming Facilities;
3. Smart Meters and the Smart Grid;[110]
4. Mobile Devices and Communications;
5. Near Field Communications (NFC);

---

[110] The Information and Privacy Commissioner of Ontario has issued a guidebook for applying the PbD concept into smart grids applications developers. Cf. [18] [VP & DK].

6. RFIDs and Sensor Technologies;
7. Redesigning IP Geolocation Data;
8. Remote Home Health Care;
9. Big Data and Data Analytics.

Since its inception in 1990s, the concept of Privacy by Design has become a key privacy protection "tool" worldwide.[111] The 32nd International Conference of Data Protection and Privacy Commissioners (Jerusalem, 2010) has recognized "Privacy by Design as an essential component of fundamental privacy protection".[112] The pending reform of the EU data protection framework has embedded a variation of PbD, i.e. data protection by design (cf. *supra*, at 2.5.5). So does the 2012 Recommendation (cf. *supra*, at 2.7.1).

2.8.3 Privacy Enhancing Technologies (PET)

Van Blarkom, Borking, and Olk, *Handbook of Privacy and Privacy Enhancing Technologies. The Case of Intelligent Software Agents* **(2003)** [19]

Privacy enhancing technologies (PET) stand for a coherent system of ICT measures that protects privacy by eliminating or reducing personal data or by preventing unnecessary and/ or undesired processing of personal data, all without losing the functionality of the information system.

European Commission, *Privacy Enhancing Technologies (PETs). The Existing Legal Framework* **(2007)**[113]

Several examples of PETs can be mentioned here.

- **Automatic anonymisation after a certain lapse of time** support the principle that the data processed should be kept in a form which permits identification of data subjects for no longer than necessary for the purposes for which the data were originally collected.
- **Encryption tools** prevent hacking when the information is transmitted over the Internet and support the data controller's obligation to take appropriate measures to protect personal data against unlawful processing.
- **Cookie-cutters blocking** cookies placed on the user's PC to make it perform certain instructions without him being aware of them, enhance compliance with the principle that data must be processed fairly and lawfully, and that the data subject must be informed about the processing going on.

---

[111] Cf. http://www.privacybydesign.ca.

[112] 32nd International Conference of Data Protection and Privacy Commissioners, *Privacy by Design Resolution*, Jerusalem, 27–29 October 2010. http://www.ipc.on.ca/site_documents/pbd-resolution.pdf.

[113] Cf. http://europa.eu/rapid/press-release_MEMO-07-159_en.pdf.

- **The Platform for Privacy Preferences (P3P)**, allowing internet users to analyze the privacy policies of websites and compare them with the user's preferences as to the information he allows to release, helps to ensure that data subjects' consent to processing of their data is an informed one.

2.8.4 Transparency Enhancing Tools (TET)

Janic, Wijbenga, and Veugen, *Transparency enhancing tools (TETs)* **(2013)** [20]

Often, data protection regulation [...] requires that users are properly informed about the fact that personal information is collected, stored, processed and disclosed, to what purpose, and how exactly, when they use a certain system. User should also be informed about third parties with which information is shared. To meet this need, the concept of Transparency Enhancing Technologies (TET) was proposed. If we define transparency as insight in how user's data is being collected, stored, processed and disclosed, TETs can then be viewed as tools providing this insight in an accurate and comprehensible way.

In contrast to PETs, these technologies exercise no particular action to enhance users privacy. They rather provide the user with necessary information on how her data is being stored, exchanged, processed and used, and as such, preserve user privacy indirectly, by enabling the user to make an informed choice on the action she finds she needs to take. [...]

Various studies report that users are reluctant to give out personal information in Internet-based transactions, due to concerns about how their personal data is being handled. It has been suggested that higher transparency by organizations on this issue might promote trust of the users and their willingness to use a particular online service.

[...] [We] define a TET as a technological tool that has one or more of the following characteristics:

- provides user with information on intended data collection, storage, processing and/or disclosure (how service providers claim to handle user's personal information)
- provides user with information on data collected, stored, processed and/or disclosed (how service providers actually handle user's personal information)
- present the above listed information in an accurate and for an average Internet user comprehensible way.

Hildebrandt, *Behavioural Biometric Profiling and Transparency Enhancing Tools* **(2009)** [21]

TETs are now defined as legal as well as technological tools. [...] This is important because we need adequate transparency rights, as instruments of constitutional democracy, as well as the technological infrastructure to be able to exercise those rights.

2.8.5 Privacy Impact Assessment (PIA)

Kloza, *Privacy Impact Assessments as a Means to Achieve the Objectives*
of Procedural Justice **(2014)** [22] (references and footnotes omitted)

A PIA is usually defined as "a process for assessing the impacts on privacy of a project, policy, programme, service, product or other initiative and, in consultation with stakeholders, for taking remedial actions as necessary in order to avoid or minimise the negative impacts" [...]. In practice, PIAs constitute a tool for supporting decision-making.

Building on the positive experience of environmental impact assessments (EIAs), the growing interest in PIAs is caused by public distrust in emerging technologies in general, by the robust development of privacy-invasive tools, by a belated public reaction against the increasingly privacy-invasive actions of both public authorities and corporations as well as by a natural development of rational management techniques [...]. From the governance viewpoint, PIAs have shifted the attention from reactive measures towards more anticipatory instruments in the belief in the rationale of an "ounce of prevention" [...]. PIAs are considered effective accountability tools that have decentralised the enforcement of privacy by focusing on the very actors involved; this is clearly visible in the 2012 European Commission's proposal for the new EU data protection framework [...].

(Recommendations for) an *ideal* PIA policy and process:

1. *Embodiment in the lifecycle of the project*: a PIA is a process that starts as early as possible (so that it can influence the design of the project), continues throughout the lifecycle of the project and is revisited afterwards, if new privacy risks are discovered. If a project "moves" to another organisation, continuity of the PIA process is ensured (i.e. PIAs as a "living instrument"). Finally, PIAs are reviewed and/or audited.

2. *Scalability*: Because organisations vary greatly in size, because the extent to which their activities intrude on privacy varies, and because their experience in dealing with privacy issues differs, organisations carry out PIAs appropriate to their own circumstances.

3. *All privacy types and beyond*: PIAs address all types of privacy and not only informational aspects thereof. If necessary, PIAs might address also ethical implications as well as issues related to surveillance.

4. *Accountability*: in the privacy protection arena, accountability not only consists of adopting and implementing the appropriate measures (i.e. the requirement of efficiency) but also in of being able to demonstrate—upon request—that such measures have been taken (i.e. the requirement of transparency) [...]. An organisation carrying out PIAs as well as assessors and senior officials are accountable for their actions and omissions related thereto. To demonstrate that the PIA process has been properly carried out and its recommendations implemented, an external audit and/or review may be conducted.

5. *Transparency*: a PIA process enjoys at least a minimum level of transparency. Both the assessors and the stakeholders must have all relevant information to assess the privacy implications of the proposed project. The requirement of transparency in PIAs is of a twofold nature: (1) of the process itself, and (2) about disclosure of relevant information, which is further split into: (a) stakeholders' participation, (b) publication of the PIA report, and (c) public registry of PIAs actually carried out. None of these precludes due respect for sensitive information.

(a) *Stakeholders' involvement*: in the PIA process, stakeholders, as representative as possible, including the public, if applicable, are identified and informed about the planned project and of the PIA process. Their views are sought and subsequently duly taken into consideration.

(b) *Report*: having concluded a PIA process, the final report is made public and is easy accessible.

(c) *Public registry*: all PIAs are listed in a public central registry, preferably in a digital form, and are easily accessible.

(d) *Sensitive information*: all these "externalities" of the PIA process, i.e. stakeholders' participation, reports and registries, beg a question about state secrets and commercially sensitive information. These are not necessarily meant to reach the public; stakeholders are usually external to an organisation carrying out PIAs. Thus, they might be consulted e.g. through closed discussion sessions with non-disclosure agreements. As far as PIA reports are concerned, an organisation could redact the documents and place confidential information in an annex and publish only the main body of the report, which is later on fed into the registry. Alternatively, an organisation might create and publish a meaningful summary of the report.

6. *Risk management and a legal compliance check*—the core elements of PIAs. Based on a proper risk management methodology, all possible risks and other negative privacy impacts are identified, assessed and—ideally—mitigated. Residual risks, if any, are justified. The assessors ensure the project's compliance with any legislative or other regulatory requirements.

7. *Internal "privacy culture"*: PIAs are only good as the process that supports them. An organisation, having set out the terms of reference of the PIA process, ensures professional and personal independence of the assessor. PIAs could be carried out in-house by e.g. a data protection officer, whose independence is sanctioned by law and by appropriate resources at her disposal (time, money, manpower) or they could be equally outsourced to an external entity whose independence is beyond any doubt. Assessors must recognise the bias and subjectivity that they might bring to the task and declare that in the report.

8. *External "privacy culture"*: PIAs needs high-level support of policy-makers, regulators and private sector. In particular, data protection authorities (DPAs) play a key role here. They promote and facilitate the PIA process by providing expertise, guidance and advice for policy-makers, organisations and assessors as well as—possibly—by reviewing and providing feedback of (selected) PIAs actually carried out.

De Hert, Kloza, and Wright, *Recommendations for a Privacy Impact Assessment Framework for the European Union* **(2012)** [23]

The key elements of the PIA process are the following [...]:

1. Determining whether a PIA is necessary (threshold analysis),
2. Identifying the PIA team and setting terms of reference,
3. Description of the proposed project,
4. Analysis of the information flows and other privacy impacts,
5. Consultation with stakeholders,
6. Risks management,
7. Legal compliance check
8. Formulation of recommendations,
9. Preparation and publication of the report,
10. Implementation of recommendations,
11. External review and/or audit,
12. Revisiting PIA if the project in question changes.

2.8.6 Legal Protection by Design (LPbD)

Hildebrandt, *Legal Protection by Design in the Smart Grid* **(2013)** [24]

LPbD insists that the legal requirements of fundamental rights such as privacy and data protection must be translated into computer system hardware, code, protocols and organisational standards to sustain the effectiveness of such right in a changing technological landscape.

1. Think in terms of data flows instead of isolated discrete data; foresee whether de-anonymisation will reinstate identifiability and treat data streams that are susceptible to such de-anonymisation as falling within the scope of data protection legislation.
2. Make privacy and security an essential part of your business-model, do not treat them as costs but as a competitive advantage—especially in the long run.
3. Start from and reiterate Data Protection Impact Assessments.
4. Practice Data Protection by Design and by Default.
5. Develop software tools and hardware infrastructure that is innovative in terms of DPbDesign and by Default.
6. Develop business models based on DPbDesign and by Default.
7. Practice Security by Design, notably end-to-end encryption and secure authentication wherever possible.
8. Invest in recurrent software analyses.
9. Practice discrimination-aware data mining.
10. Base your trust management on trustworthiness.
11. Never underestimate the recurrent cost of safety and security.
12. Don't allow critical infrastructure to depend on volatile markets.
13. Create separate data streams for (1) critical infrastructure that protects the right to universal service, and (2) commercial value added services.
14. Design profile transparency in the back-end of the Smart Grid system.
15. Design intuitive interfaces that provide transparency about the potential consequences of sharing one's data (showing what profiles they match).
16. Design for profile transparency in the front-end of the Smart Grid system (allow consumers to play around with their data to figure out how they are matched).

2.8.7 Privacy Certification

Privacy certification is a relatively new addition to the EU data protection field. Although the 1995 Data Protection Directive makes no explicit mention to it, neither are its provisions particularly accommodating to its special needs and requirements, the notion of using straightforward identification means in order for individuals to quickly be able to distinguish sound data protection practices has expanded within EU Member States over the past years. Today, a multitude of national privacy certification schemes are in place—initiatives, however, remain fragmented and piecemeal, undermined both by lack of general data protection harmonization across the EU and lack of formal recognition by the law or any other competent (state) authority.[114]

However, because the significance of privacy certification while providing individuals with simple means to make quick decisions, and therefore assist commerce and processing technologies, was formally acknowledged by the European Commission,[115] special mention to it was made in the original draft of the General Data Protection Regulation.[116] Its Art 39(1) sets that:

> 1. The Member States and the Commission shall encourage, in particular at European level, the establishment of data protection certification mechanisms and of data protection seals and marks, allowing data subjects to quickly assess the level of data protection provided by controllers and processors. The data protection certifications mechanisms shall contribute to the proper application of this Regulation, taking account of the specific features of the various sectors and different processing operations.

This approach was substantially altered by the European Parliament input[117] and should therefore not be perceived as final (given also the long negotiation period that lies ahead before the GDPR comes into effect). What could be taken for granted, however, is the fact that privacy certification will in some way be present in its final text, as an important accessory to the data protection purposes. Such privacy certification will evidently be industry-specific: although the distinction whether products and/or services may be certified needs to be finalized, smart grid technologies and systems, if not also services, constitute an obvious candidate. Depending on the particulars of the certification implementation that will ultimately be adopted (for instance, whether it will be EU or Member State run, who will the accreditors be, what will be the role of the industry concerned) smart grid participants might find themselves involved in different ways in the relevant processes before having to actually accredit their own personal data processing. Whichever the case may be, this should be a welcome development for them: privacy certification, if properly implemented, warrants legal certainty to data controllers entering a new personal data processing field while also fostering public trust and awareness—both necessary characteristics for this field to expand.

---

[114] European Commission, Joint Research Centre, *EU privacy seals project, Inventory and analysis of privacy certification schemes*, 2013.

[115] European Commission, *A comprehensive approach on personal data protection in the European Union*, COM (2010) 609 final, p. 12.

[116] Cf. *supra*, note 75.

[117] European Parliament, Committee on Civil Liberties, Justice and Home Affairs, *Report on the proposal for a regulation of the European Parliament and of the Council on the protection of individuals with regard to the processing of personal data and on the free movement of such data (General Data Protection Regulation)*, 21 November 2013. http://www.europarl.europa.eu/sides/getDoc.do?pubRef=-//EP//NONSGML+REPORT+A7-2013-0402+0+DOC+PDF+V0//EN.

2.8.8 Overview of Applicable Privacy-Friendly Algorithms for Smart Metering

As a response to the growing need to ensure adequate protection of privacy and personal data in smart grid and smart metering systems, in recent years developed have been a number of privacy-friendly algorithms for processing information in these systems.[118]

2.8.8.1 Microsoft Research: Privacy-Friendly Smart Metering

Danezis and Rial, *Privacy-Preserving Metering for Smart-Grids* (2010) [25][119]

We propose a secure protocol between a customer's electricity meter and a utility provider that reveals the total consumption fee, while keeping private the individual measurements. Furthermore, it guarantees the fee is correct and derived according to the fine-grain meter measurements and the tariff policy of the provider.

In our system the utility provider sets a tariff policy, signs it and sends it to the user. The pricing policy maps consumption and other parameters such as the time of day, to a tariff. Over the billing period, meters output readings and other metering information and sign them. Periodically, the user, their device or a service of their choice, uses the readings to compute a payment message that includes the total fee and a mathematical proof that the fee is correct, given the applicable tariffs and the readings. Upon receiving the payment message the provider can check the correctness of the bill without learning any information about the individual meter readings.

Additionally, our protocol allows for the selective disclosure of consumption data to the provider, but only with the users' consent. The selective disclosure of fine-grained data is certified to be correct, and other function of readings can be computed and revealed. Besides the final bill other information can be released, such as the total consumption within the billing period, as it is done today, to facilitate network management and fraud prevention.

Security for the provider relies on established signature schemes and the binding property of cryptographic commitments. User privacy relies on the zero-knowledge property of proofs of knowledge and on the hiding property of commitments. The provider ensures that the signatures on the meter readings are correct though tamper-evident hardware as for conventional metering security. Our protocols are proved secure using well established cryptographic techniques.

Users can delegate the calculation of their bill to any device or service they wish […]: a home server, an on-line service or a smart meter. No matter what their choice is, the provider and everyone else can verify in case of dispute that the final bill is correct, and does not need to trust the actual computation or the party that performed it. The freedom to perform the computation of the bill locally or through any device or delegate, without revealing the detailed readings, allows the user to preserve their privacy.

---

[118] We thank Michael John for pointing this to our attention.

[119] This algorithm has been discussed in a greater detail, among others, in [26, 1–12], [27, 175–191] and [28, 148–162].

Our technology combines the benefits of fine grained metering and charging with users' needs for privacy. It is applicable to a number of settings beyond utility metering, such as pay-as-you-drive car insurance, road tolling and taxation, utility billing or software licence management, providing integrity and privacy to any billing process.

2.8.8.2 Privacy-Friendly Energy Metering via Homomorphic Encryption

Garcia and Jacobs, *Security and Trust Management* (2011) [29]

The first part of the paper discusses general issues in (electricity) metering and argues towards the inclusion of a trusted element, like a smart card, in E-meters. This is reflected in the slogan "power to the meter!". Such a trusted element provides secure storage of meter readings (like the traditional meter does via hardware protection), and basic cryptographic primitives based on public key cryptography, for authentication and secure communication. The protocols later on in the paper are based on the availability of such primitives. They demonstrate how basic cryptographic techniques can be used to achieve justifiable monitoring aims of grid operators without violating privacy of consumers.

In particular, Sect. 4 describes a protocol whereby data concentrators at the neighbourhood level can obtain sums of the measurements of all the connected customers (typically a few hundred) without learning the individual measurements. By comparing this sum with its own measurement of the consumed amount, it becomes clear how much energy leaks in this neighboorhood. These protocols may be run frequently, say every 15 min, without affecting privacy.

2.8.8.3 I Have a DREAM! (DiffeRentially PrivatE SmArt Metering)

Acs and Castelluccia, *I have a DREAM! (DiffeRentially PrivatE SmArt Metering)* (2011) [30]

With our scheme, an (electricity) supplier can periodically collect data from smart meters and derive aggregated statistics without learning anything about the activities of individual households. For example, a supplier cannot tell from a user's trace whether or when he watched TV or turned on heating. Our scheme is simple, efficient and practical. Processing cost is very limited: smart meters only have to add noise to their data and encrypt the results with an efficient stream cipher.

2.8.8.4 TRUST: Team for Research in Ubiquitous Secure Technology

Lisovich and Wicker, *Privacy Concerns in Upcoming Residential and Commercial Demand-Response Systems* (2008) [1] (references omitted)

VII. ALGORITHM ROBUSTNESS AND PRIVACY SOLUTIONS

We would like to comment on the algorithm's robustness, and by extension on the informational content contained within the data. We do this by measuring the effect of increased data granularity on the estimation of presence intervals.

Although what follows is a comment rather than a complete analysis, the parameter's tolerance to data scarcity gives an upper bound on the dataset's informational content and provides sufficient ground for us to discuss the relationship between data granularity and privacy solutions.

We believe that privacy protection ultimately lies in policy. However, it's worthwhile to examine technological solutions. Privacy can be preserved through technological means by decreasing the data's information content through signal processing. Such processing may form a useful part of a policy solution—interested parties may be given lower resolution data (resolution depending on its intended use) as a way of ensuring their compliance with stated privacy policies. Additionally, consumers may choose to control the amount of information content leaving their home (in this case, the signal processing is performed in-residence by the meters), exchanging quality of service for privacy protection.

There are several ways to increase the granularity of data. The original dataset can be passed through a filter, downsampled, or corrupted by noise. In particular, a lowpass filter may be applied to remove events of high frequency, masking events which rapidly trigger between 'on' and 'off' states. No matter what is done to the high-resolution data, it is important to retain weekly/monthly electricity usage numbers, since the data's analysts will want true averages and totals for billing and research purposes. […]

IX. GUIDELINES

A report to the California Energy Commission, written in part by our Berkeley colleagues, makes several recommendations for power-data handling. They recommend:

1. Multiple tiers of control and oversight, both by the utilities themselves and the state/ federal government.
2. Explicit guidelines regulating access to data for customer service, load research, and other functions.
3. Strong user control over information leaving the residence.
4. Protocols which do most of the data processing at stations located inside the residence, as well hard prohibitions against relaying certain types of data.

2.8.8.5 Elster's Algorithm

Elster, *Privacy Enhancing Technologies for the Smart Grid. Elster's Proposal for Privacy Enhancing Technology Implementation* (2012) [31]

[…] for the Smart Grid most use cases for grid maintenance do not need individual customer data, but an aggregate of the data generated by a number of customers. In many cases, the only reason why individual data is collected is that the aggregated data cannot be measured directly. The easiest way to get to the data that is truly required, the aggregates, is to collect and combine individual data.

The implemented PETs use homomorphic cryptographic methods that enable computations on encrypted input data to be performed, but reveal only the result of the computation in unencrypted form. The cryptography used is optimized specifically for the task of aggregating data from several sources, which allows for an extremely compact and efficient implementation.

In the Smart Grid scenario, this means that any Smart Meters involved in the process offer measurement data in an encrypted form that is unusable until aggregated with measurements from other meters.

If the meters have pair wise shared secret keys, this can be done by one group of meters generating a set of values that add up to zero, which are then used to mask the real measurements. The advantage of this method is that it has a very low computational overhead, as the masking values can easily be calculated from the shared keys, and no communication overhead at all. To establish the keys required, a variant of the well-established Diffie–Hellman Key exchange protocol that requires each meter to only use one single public/private key pair is applied.

As the aggregate cannot be computed if individual measurements are missing, the group size should be kept reasonably small, ideally up to twenty meters. Larger groups can then easily be generated by adding the measurements of smaller subgroups, with no risk of endangering of the privacy properties. […]

In order to compute an aggregate using traditional readings, each measurement value needs to be decrypted individually. By doing so, the individual consumption from each meter— and thus from each household—is revealed.

This is not the case with a privacy enhancing reading. […] In this case, the meter readings are encrypted and stored in the privacy enhancing load profile, and the aggregated sum can be computed over the encrypted readings without needing to decrypt individual values.

Individual values are never obtained and no personal information is transferred with the reading. Comparing the computed sums of the traditional and the privacy enhancing reading demonstrates that the protocol is correct since both are always identical.

2.8.8.6 Data Aggregation Protocols DiPA (Diffie–Hellman-Based Private Aggregation) and LoPA (Low Overhead Private Aggregation Protocol)

Kursawe, *How to have the cake and Eat it, too: Protecting Privacy and Energy Efficiency in the Smart Grid* (2012) [32] (reference omitted)

4.2. The Data Aggregation protocols DiPA and LoPA

In this section, we outline two concrete protocols for privacy protecting data aggregation. For the scope of this work, we only give an intuition on the protocol mechanism and a basic outline [...].

The first protocol, DiPA (Diffie–Hellman based Private Aggregation) is a simple cryptographic protocol based on the Diffie–Hellman public key scheme. The main mechanism here is a *homomorphic commitment scheme*, which can be implemented very efficiently using Diffie–Hellman on elliptic curves.

A commitment scheme is a simpler tool than an encryption scheme. It allows a user to fix some secret (*commit* to it), and to later reveal the secret and prove that this was the value she committed to. As opposed to an encryption, a commitment scheme is easier to implement, and it does not need a secret decryption key, which means that there is less key management required.

As visualization, one can think of the original value as a Lego-car, the commitment scheme as a rubber hammer, and the commitment as a heap of Lego-bricks. Committing to a value (car) means to smash it with the hammer, and showing the heap of stones to the verifier. It is now computationally hard for the verifier to reconstruct the car from the heap, while it is easy to verify that a given car corresponds to the given heap.

The special property of *homomorphic commitments* is that it is possible to perform computations on the commitments, which then correspond to computations on the original plaintext, i.e.,

$$\text{Commit}(A + B) = \text{Commit}(A) * \text{Commit}(B)$$

In our visualization, this means that two Lego cars can be added up to a transformer (which is the addition on the original value side). Similarly, if one adds the heaps generated by the two individual cars, one gets the heap generated by the transformer, so the addition on the original values has an equivalent operation on the commitments. [...]

While this protocol only allows us to compare values we already know, it is easy to transform into a protocol that computes actual aggregates. To this end, we simply perform the standard aggregation protocol on individual bytes, and then brute force those on the backend server (given the small domain of measurement values, this should not be more than a few hundred tests, which a modern PC can easily handle).

A main advantage of this protocol is that it allows for different sets of meters to be aggregated on, without requiring any change to the meter configuration. Instead, the aggregator needs to know the sum of the corresponding masking values. This would allow, for example, to separately aggregate over all meters in one particular district, as well as over all meters of consumers that also generate energy.

The LoPA (Low overhead Private Aggregation protocol) is even simpler, but does sacrifice some flexibility for this simplicity. In this protocol, the group of meters whose measurements are aggregated is fixed, and all meters in one group know of each other. Each two meters in one aggregation group share a common secret x (i.e., in a group of ten meters, each meter needs to keep nine such secrets). When a measurement is to be protected, one meter adds its corresponding x for all its peers to its output value, while the other one subtracts it. Thus, the overall effect of the secrets cancels out completely, and an aggregator summing up all values gets the exact sum of all measurements. However, if only one measurement is missing, not all the secrets cancel out, and the reading is unreadable. To protect privacy over several readings, the secrets need to be changed after each reading; this can easily be done without any interaction and little computational overhead by applying a hash function such as SHA-256. [...]

One major advantage of this approach is that there is no public key cryptography involved once the system is initialized, and all operations are simple additions as well as a hash function. This not only reduces the computation overhead to the absolute minimum, but also allows the message size to stay exactly the same—a masked 32-bit value still is a 32-bit value, as opposed to the homomorphic commitment based protocol, where it needs to be long enough to be cryptographically secure. Thus, this protocol integrates very neatly into the existing DLMS/COSEM standard, and no changes have to be made to the message format.

The price is a somewhat a smaller flexibility—in this protocol the aggregation group is fixed by the keys the meters have, and it is not possible to aggregate over different sets of meters simultaneously. Also, a meter does need enough memory to store all the shared keys with its peers. This does not have a large impact in practice, however, as the sets of meters should be kept small anyhow for stability reasons. [...]

6. Conclusions

[...] The main message is twofold. Firstly, modern privacy enhancing technologies have reached a level of practicability that does allow them to work in a real system—and while larger scale tests still have to be done before a real deployment is possible, the implementation already demonstrates that an integration into existing architectures and hardware is feasible. Secondly, we show a practical example of 'positive-sum' privacy, i.e., a privacy technology that has been developed together with the businesses, and that does fit into the overall business model and its requirements. In doing so, the technology even can generate positive value for the business—not only by helping to comply to regulation and saving costs on otherwise needed technology, but by allowing to have more privacy and actually use more data.

2.8.9 Going Beyond Mere Privacy: Technology Assessment (TA)

Est and Brom, *Technology Assessment, Analytic and Democratic Practice* **(2012)** [33] (reference omitted)

Argumentative TA is a mode of TA that wants to deepen the political and normative debate about science, technology, and society. Classical TA is a form of expert-based policy analysis to identify and evaluate in an early stage the potential secondary consequences of technology.

Constructive TA is a mode of TA that wants to address social issues around technology by influencing design practices.

Parliamentary TA […] (aims) to strengthen representative democracy by timely informing MPs about the potential social impacts of technological change.

Participatory TA is a mode of TA that aims to enrich the political and public debate around the social aspects of science and technology by organizing the involvement of experts, stakeholders, and citizens to identify and evaluate the societal impact of technological change.

Technology assessment (TA) is a scientific, interactive, and communicative process that aims to contribute to the formation of public and political opinion on societal aspects of science and technology.

Introduction

Technology assessment (TA) deals with the relationship between technological change and social problems. In essence, therefore, TA has a strong political dimension to it. There is a core belief that drives TA. One could speak about a philosophy of TA, which became widely accepted during the 1970s and 1980s throughout the United States and Europe, and which is gradually impacting Southern America and Asia as well. Rip articulates this philosophy of TA as follows:

to reduce the social costs of learning by error, and to do so by systematic anticipation of potential impacts of new technologies and large projects, and feedback into decision making.

In other words, TA combines an awareness about potential negative and positive effects of technological change with the belief or hope that one can anticipate these effects.

2.9 Consumer Empowerment

Bureau Européen des Unions de Consommateurs, *Protecting and Empowering Consumers in Future Smart Energy Markets* **(2013)** [34][120]

The retail energy market is often perceived by consumers as being rather complex and future smart energy markets will pose even bigger challenges for them. In a world of fast paced technological innovation and changes, consumers may face yet other new technologies to enter their everyday life. And although these new technologies may offer new services for consumers, the benefits for them are not yet guaranteed. In well-functioning retail energy markets, consumers must be informed and sufficiently protected so that they can benefit from competition, compare information on consumption and costs, and know their rights and means of dispute resolution. Any technological development should ensure user-friendliness and consumer engagement built on consumer protection and empowerment.

---

[120] Further readings include [35, 36].

In this position paper, BEUC, The European Consumer Organisation, points out to challenges ahead and outlines all necessary elements which need to be addressed to ensure that European consumers are well-protected and empowered in future smart energy markets.

Installation of Smart Meters and Roll-Out Strategies

- Cost-Benefit Analyses must be mandatory and take into account the impact on different consumer groups.
- Cost-Benefit Analyses must be carried out by an independent organisation and regularly reviewed during the smart meter roll-out.
- The European Commission should undertake a continuous assessment of national roll-outs.
- Consumers should freely choose if they want to use smart meters in their homes. This is particularly important where consumers are required to pay for it.
- When consumers do not accept a smart meter, they should not bear any additional costs.
- A coordinated bottom-up approach could facilitate the provision of more targeted information about the potential benefits and new services available to consumers.
- Member States should outline a clear vision of the benefits for consumers and a strategy on how these will be delivered.
- Member States must report annually on the achieved progress and the costs and benefits for consumers.
- Policy makers must provide for a solid legal and regulatory framework that guarantees that the smart meter roll-out is cost efficient and costs and benefits are fairly shared among all stakeholders that benefit from the new technology.
- Targeted information and personalised advice are necessary to raise awareness about how consumers can achieve potential benefits from smart meters and related services.
- Consumers must be provided with on-going advice and support during and after the installation of smart meters.
- During the installation visit, consumers should not be sold any tariffs, goods or services.
- Measures protecting vulnerable consumers must be in place so that the roll-out of smart metering technology does properly address their situation.

Potential benefits for consumers

- Consumers equipped with smart meters should get accurate and regular bills based on actual consumption.
- Functionalities of smart meters must enable access to real time information as well as historical information, advice and easy switch.
- Member States should analyse and present evidence of what frequency of historical consumption data works for consumers.
- If Member States allow the use of the remote disconnection functionality, they must put in place safeguards and legal protection so that this functionality cannot be misused.
- Distributional analysis must be performed on the impact of time-of-use tariffs on different social groups and if/how these groups can access the benefits of new deals.

- Demand response programmes should be available to consumers on opt-in basis.
- As smart meters will enable dynamic pricing tariffs, National Regulatory Authorities must carefully monitor the tariff complexity and ensure new tariffs are easy to compare and do not prevent switching.
- Clear information and protection frameworks about best use, remote control and disconnection of smart appliances must be provided to consumers.
- Those consumers who are also producers should receive information in an appropriate format so that they understand the full potential of micro-generation.

Technological Aspects of Smart Metering Systems

- Interoperability and modularity of the system must be ensured in order to avoid lock- ins and ensure the system is future-proof.
- The technology should meet inclusivity by design standards to ensure consumers find it easy to use.
- Reliability and quality of energy supply need to be monitored.
- Consumers need information to be easily accessible.

Protection of consumers' personal data

- Processing of personal data must be fair, lawful and must comply with the principles of data protection processing, including transparency, data minimisation and purpose limitation.
- Each processing operation must be based on the most appropriate legal ground, while the legitimate interests of the data controller should not be used as a loophole.
- Compliance with the principles of privacy by design and privacy by default must be ensured.
- Introduction of mandatory Data Protection and Privacy Impact Assessment (DPIA/PIA) should be conducted on all aspects of smart metering.
- Retention of personal data should not exceed what is absolutely necessary for specific and lawful purpose.
- Consumers' personal data should be stored at the consumer's side by default.
- Effective enforcement of the Data Protection legislation is key.

2.10 Case Studies

2.10.1 The Netherlands

Cuijpers and Koops, *Smart Metering and Privacy in Europe:*
***Lessons from the Dutch Case* (2013)** [37] (footnotes omitted)

Rolling out smart meters, however, requires smart legislation. The Dutch case, where the Senate blocked two smart metering bills in 2009, demonstrates that introducing smart meters can be significantly delayed if the underlying legislation if flawed.

More in particular, the Dutch case shows that privacy is not to be underestimated. The failure of doing an *ex ante* privacy impact assessment backfired, as the proposed laws required mandatory installation in every household of smart meters that would send quarter-hourly/hourly measurements to network operators and daily measurements to energy suppliers. This level of detail creates privacy-sensitive data, and the necessity of smart meters infringing people's privacy in this way had not been substantiated by the government.

Several lessons can be learned from the Dutch case for countries considering smart metering legislation. In terms of substance, the level of detail of smart meter readings and the mandatory or voluntary character of smart meters are crucial issues to take into account. In terms of procedure, a privacy impact assessment is vital to identify at an early stage the potential effects on individuals' privacy and to choose the least privacy-infringing modalities of smart metering. Pitfalls of function creep should be avoided by resisting the temptation of making a meter 'too smart' all at once, which could easily lead, as the Dutch case demonstrates, to choosing privacy-invasive instead of privacy-friendly settings; such settings are unnecessary to achieve the primary purpose of the current European energy-efficiency regulation, namely to provide consumers with sufficient feedback on their energy consumption to induce energy-saving behaviour.

The procedural lessons also highlight the need for privacy by design. This principle concerns the need to integrate, at practical level, data protection and privacy from the very inception of new information and communication technologies. The purpose, design, functionalities and implementation of the smart metering system determines to a large extent whether or not it will comply with privacy and data protection legislation. Therefore, from the beginning, privacy and data protection law must be taken into account as an important requirement for the design of smart metering systems. It is a promising development that the proposed Regulation on data protection explicitly establishes obligations for privacy by design and default, and an *ex ante* obligation for data protection impact assessments in cases where data processing has specific risks.

The substantive lessons can also be formulated in the form of a key trade-off for legislators: the 'smartness' of the meter versus a comprehensive, mandatory roll-out. The smarter a meter is, i.e., the more detailed its readings are—up to quarter-hourly or even less—and the more functionalities it has, the more likely is it to be privacy-invasive. Current research already shows how revealing smart meter data can be of people's daily life in their homes, and findings such as the capacity to derive which TV channel one is watching from real-time energy readings suggest that the privacy-sensitivity of energy consumption data will only increase in the future. This implies that if countries opt for smart meters with detailed readings that leave the privacy of the home, this can hardly be considered necessary in a democratic society, and hence, such smart meters can only be rolled out on a voluntary basis, as now will happen in the Netherlands. And conversely, if countries choose a relatively 'dumb' meter that conforms to the minimum requirements of European legislation (capable of at least daily measurements and with an interface showing readings to the customer), they can likely make the roll-out of such meters mandatory for consumers, in terms of compliance with Art 8 ECHR.

2.10.2 The United Kingdom

Brown, *Britain's Smart Meter Programme: A Case Study in Privacy by Design* **(2013)** [38]

The British programme proposed to Parliament by DECC[121] has ultimately ended up with similar rules to the amended Dutch programme: meter installation is voluntary for customers; energy consumption is measured for billing purposes without specific consent at monthly (Britain) or bimonthly (Netherlands) intervals, and when customers move or change suppliers. More detailed data can be read for specific legal obligations, but explicit consent is needed for half-hourly (Britain) or hourly (Netherlands) readings to be taken for other purposes.

While these compromises seem to meet the basic requirements of the Data Protection Directive and European Convention on Human Rights, earlier consideration of more privacy-friendly options might have produced a more protective (and cheaper) system. For example, Ross Anderson has suggested that governments should simply coordinate the production of industry standards for communication between meters, home devices, and energy companies, and set privacy standards clarifying that data should be controlled by consumers on the meter and shared minimally with other parties. This would allow privacy-sensitive consumers to use privacy-enhancing technologies such as those developed by Danezis et al.,[122] while still providing the energy industry with the information it needs to manage networks and supplies. Unfortunately, the British government now appears to have become too attached to their current proposals to reconsider this option.

2.11 Observations: Key Points

1. The deployment of smart grid and smart metering systems, as a highly invasive surveillance tool, has raised serious privacy, data protection, and other ethical concerns. These were, to some extent, subsequently taken into consideration by the EU regulator. This is particularly important since the EU Third Energy Package (2009) made the roll-out of smart metering compulsory with a goal to achieve 80 % of deployment by 2020, should the economic assessment be positive.

2. Since any smart grid and smart metering system is capable of processing personal data, the general data protection framework applies. Such a system raises a number of data protection issues, such as the classification of data processed within smart metering as personal or technical data, the distinction between the data processors and controllers, the application of the data minimization principle, the length of data retention, the scope and exercise of the data subject's rights, the legal basis for processing, as well as security and confidentiality of data processing. However, exact solutions to these issues highly depend on the technical design of a given smart grid solution [39, 196].

---

[121] Department of Energy and Climate Change [VP & DK].

[122] Cf. *supra*, at 2.8.8.1 [VP & DK].

3. The general data protection principles, embodied in particular in the 1995 Data Protection Directive, proved to be sufficiently clear and satisfactory, but there is a need for tailoring them down to a more concrete regulatory level [39, 196].
4. There is no specific binding legislation in the EU devoted to privacy, data protection, and security issues in smart grids and smart metering systems. Furthermore, as of the time of writing, there is no specific case law of senior European Courts devoted thereto. In contrast, a number of soft law (i.e. non-binding) legal instruments have been promulgated to supplement the binding EU data protection framework and thus achieve the desired policy goals. However, the efficiency of the latter—due to their voluntary nature—is questionable.
5. It is fair to say that the main "tool" for the protection of privacy and personal data is a data protection impact assessment (DPIA). To that end, a template has been developed. The pending reform of the EU data protection framework might make a DPIA compulsory. Further "tools" include data protection by default and data protection by design.
6. Privacy and data protection issues are a necessity that all policy makers and actors take into consideration. In particular, the recent Dutch and British case studies give an example of balancing the energy policy goals with privacy and data protection concerns.

References

1. Lisovich MA, Wicker SB (2008) Privacy concerns in upcoming residential and commercial demand-response systems. In: 2008 Clemson University Power Systems Conference, vol 1, pp 1–10, https://www.truststc.org/pubs/332.html
2. Anderson R, Fuloria S (2010) Who controls the off switch? In: Proceedings of the IEEE SmartGridComm, pp 170–190 http://www.cl.cam.ac.uk/*rja14/Papers/meters-offswitch.pdf
3. Department of Energy and Climate Change (2012) Smart metering implementation programme: data access and privacy: government response to consultation. London, December 2012, https://www.gov.uk/government/uploads/system/uploads/attachment_data/file/43046/7225-gov-resp-sm-data-access-privacy.pdf
4. Smart Grid Interoperability Panel—Cyber Security Working Group (2010) Guidelines for smart grid cyber security: privacy and the smart grid. NIST, vol. 2, August 2010. http://csrc.nist.gov/publications/nistir/ir7628/nistir-7628_vol2.pdf
5. Craig PP, de Búrca G (2008) EU law: text cases and materials. OUP, Oxford
6. Nugent N (2010) The government and politics of the European Union, 7th edn. Palgrave Macmillan, Basingstoke
7. Cini M, Borragan NPS (2010) European union politics. OUP, Oxford
8. Wacks R (2010) Privacy: a very short introduction. OUP, Oxford, pp 30–31
9. Brandeis LD, Warren S (1890) The right to privacy. Harvard L Rev 4:193
10. European Union Agency for Fundamental Rights (2013) Handbook on European data protection law. Vienna, Strasbourg. doi: 10.2811/53711
11. Court of Justice of the European (2014) The court of justice declares the data retention directive to be invalid. Press Release (54/14):2, http://curia.europa.eu/jcms/upload/docs/application/pdf/2014-04/cp140054en.pdf

12. Murrill BJ, Liu EC, Thompson II RM (2012) Smart meter data: privacy and security. Congressional Research Service, 3 Feb 2012, https://fas.org/sgp/crs/misc/R42338.pdf
13. Knapp E, Samani R (2013) Applied cyber security and the smart grid. Elsevier, London, pp 88ff
14. Bleicher A (2010) Privacy on the Smart Grid. Are smart meters spies? They don't have to be. IEEE Spectrum, http://spectrum.ieee.org/energy/the-smarter-grid/privacy-on-the-smart-grid
15. Spiekermann S (2012) The RFID PIA—developed by industry, endorsed by regulators. In: Wright D, De Hert P (eds) Privacy impact assessment, pp 323–346. doi: 10.1007/978-94-007-2543-0_15
16. Cavoukian A (2012) Operationalizing privacy by design: a guide to implementing strong privacy practices, pp 21–25
17. Cavoukian A (2013a) Privacy by design. Toronto, pp 1–6
18. Cavoukian A (2013b) Privacy by design: fundamentals for smart grid app developers. Toronto
19. Van Blarkom GW, Borking J, Olk J (2003) Handbook of privacy and privacy-enhancing technologies. The case of intelligent software agents. College Bescherming Persoonsgegevens, The Hague, p 33, http://www.andrewpatrick.ca/pisa/handbook/Handbook_Privacy_and_PET_final.pdf
20. Janic M, Wijbenga JP, Veugen T (2013) Transparency enhancing tools (TETs): an overview. In: 2013 third workshop on socio-technical aspects in security and trust, June, pp 18–25. doi: 10.1109/STAST.2013.11
21. Hildebrandt M (2009) Behavioural biometric profiling and transparency enhancing tools. Deliverable D7.12 of the "Future of identity in the information society" project [FIDIS], p 20, http://www.fidis.net/fileadmin/fidis/deliverables/fidis-wp7-del7.12_behavioural-biometric_profiling_and_transparency_enhancing_tools.pdf
22. Kloza D (2014) Privacy impact assessments as a means to achieve the objectives of procedural justice. In: Schweighofer E, Kummer F, Hötzendorfer W (eds) Transparenz. Tagungsband Des 17. Internationeln Rechtsinformatik Symposions IRIS 2014. Osterreichische Computer Gesellschaft, Vienna, pp 449–458
23. De Hert P, Kloza D, Wright D (2012) Recommendations for a privacy impact assessment framework for the European Union. Brussels, London, pp 12–13, http://piafproject.eu/ref/PIAF_D3_final.pdf
24. Hildebrandt M (2013) Legal protection by design in the smart grid. Privacy, data protection and profile transparency. Smart Energy Collective, Arnhem, http://pilab.nl/wp-content/uploads/2013/05/KEM-64P707-BRO-LPbD-in-SmartGrid_A4_FC_v4.pdf
25. Danezis G, Rial A (2010) Privacy-preserving metering for smart-grids. Executive summary, p 1, http://research.microsoft.com/en-us/projects/privacy_in_metering/
26. Rial A, Danezis G (2011) Privacy-preserving smart metering. In: Proceedings of the 10th annual ACM workshop on Privacy in the electronic society, ACM, pp 49–60. doi: 10.1145/2046556.2046564
27. Kursawe K, Danezis G, Kohlweiss M (2011) Privacy-friendly aggregation for the smart-grid. In: PETS'11 proceedings of the 11th international conference on privacy enhancing technologies. Springer, Berlin, Heidelberg, pp 175–191. doi: 10.1007/978-3-642-22263-4_10
28. Danezis G, Kohlweiss M, Rial A (2011) Differentially private billing with rebates. In: Information hiding. Springer, Berlin, pp 148–162. doi: 10.1007/978-3-642-24178-9_11
29. Garcia FD, Jacobs B (2011) Privacy-friendly energy-metering via homomorphic encryption. In: Security and trust management. Springer, Berlin, pp 226–238. doi: 10.1007/978-3-642-22444-7_15
30. Acs G, Castelluccia C (2011) I have a DREAM! (DiffeRentially privatE smArt Metering). In: Information hiding. Springer, Berlin, pp 118–132. doi: 10.1007/978-3-642-24178-9_9
31. Elster (2012) Privacy enhancing technologies for the smart grid. Elster's proposal for privacy enhancing technology implementation, pp 1–8, http://www.elster.com/assets/downloads/PETwhitePaperA4-Web.pdf
32. Kursawe K (2012) How to have the cake and eat it, too: protecting privacy and energy efficiency in the smart grid. In: Pohlmann N, Reimer H, Schneider W (eds) Securing electronic

business processes: highlights of the information security solutions Europe 2011 conference. Vieweg + Teubner, Wiesbaden, pp 164–173

33. Est RV, Brom F (2012) Technology assessment, analytic and democratic practice. In: Encyclopedia of applied ethics, 2nd edn, pp 306–320. doi: 10.1016/B978-0-12-373932-2. 00010-7

34. Bureau Européen des Unions de Consommateurs (2013) Protecting and empowering consumers in future smart energy markets. Brussels, pp 3–4, http://www.beuc.org/publications/2013-00083-01-e.pdf

35. Klopfert F, Wallenborn G (2011) Empowering consumers through smart metering. Bureau Européen des Unions de Consommateurs, Bruxelles, http://www.beuc.org/publications/2012-00369-01-e.pdf

36. Hoenkamp R, Huitema GB, de Moor-van Vugt AJC (2011) The neglected consumer: the case of the smart meter rollout in the Netherlands. Renew Energy Law Policy (RELP) 4:269–282

37. Cuijpers C, Koops B-J (2013) Smart metering and privacy in Europe: lessons from the Dutch case. In: Gutwirth S, Leenes R, De Hert P, Poullet Y (eds) European data protection: coming of age, pp 269–293. doi: 10.1007/978-94-007-5170-5_12

38. Brown I (2013) Britain's smart meter programme: a case study in privacy by design. Int Rev Law Comput Technol 28(2):172–184. doi: 10.1080/13600869.2013.801580

39. De Hert P, Kloza D (2011) The challenges to privacy and data protection posed by smart grids. In: Schweighofer E, Kummer F (eds) Europäische Projektkultur Als Beitrag Zur Rationalisierung Des Rechts. Tagungsband Des 14. Internationalen Rechtsinformatik Symposions IRIS 2011. Osterreichische Computer Gesellschaft, pp 191–196

Printed in the United States
By Bookmasters